Zinc and Zirconia

ROYAL SOCIETY OF CHEMISTRY

Zinc and zirconia

Edited by John Johnston, Ted Lister and Colin Osborne

Designed by Imogen Bertin and Sara Roberts

Published by the Education Division, The Royal Society of Chemistry

Printed by The Royal Society of Chemistry

Copyright © The Royal Society of Chemistry 1999

Apart from any fair dealing for the purposes of research or private study, or criticism or review, as permitted under the UK Copyright Designs and Patents Act, 1988, this publication may not be reproduced, stored, or transmitted, in any form or by any means, without the prior permission in writing of the publishers, or in the case of reprographic reproduction, only in accordance with the terms of the licences issued by the Copyright Licensing Agency in the UK, or in accordance with the terms of licenses issued by the appropriate Reproduction Rights Organization outside the UK. Enquiries concerning reproduction outside the terms stated here should be sent to The Royal Society of Chemistry at the London address printed on this page.

For further information on other educational activities undertaken by The Royal Society of Chemistry write to:

The Education Department
The Royal Society of Chemistry
Burlington House
Piccadilly
London W1V 0BN

Information on other Royal Society of Chemistry activities can be found on its websites:
www.rsc.org
www.chemsoc.org

ISBN 0 85404 955 X

British Library Cataloguing in Data.
A catalogue for this book is available from the British Library.

Pictures used in the sections "Zinc – ancient mystery, modern marvel" and "Zinc – for the roof over your head", and the top three pictures on the back cover were supplied by Spectrum Colour Library. The bottom three pictures on the back cover are reproduced by kind permission of Dynamic Ceramics tel: 01270 501000. The front cover picture is reproduced by kind permission of Britannia Zinc.

Zinc

General introduction

The material in this book is the result of a Learning Material Workshop organised by The Royal Society of Chemistry, The Worshipful Company of Armourers and Brasiers and The Institute of Materials. Two workshops were held at:

- Britannia Zinc, Avonmouth, on the extraction of zinc; and
- MEL Chemicals, Bolton, on the manufacture of zirconia.

A different group of chemistry teachers was involved in each workshop. The groups spent a day with one of the companies and was given a presentation on the plant. The following day was spent brainstorming and drafting the material which is presented here in edited form.

The book contains teacher's notes and material to photocopy for use of students.

Contents

ZINC AND ITS EXTRACTION

Introduction 3
The structure of the material 4

Part 1 Pre-16

Teacher's notes...6
Answers..7
Worksheets
Information sheet *Zinc at Avonmouth*
Comprehension exercise *Zinc smelting at Avonmouth*
Comprehension exercise *The extraction of zinc*
Further worksheets
Zinc – ancient mystery, modern marvel
Tutty – the metal that turns copper into gold
In the beginning, there was… ZINC
William Champion – Zinc Smelter
From horizontal to vertical – zinc smelting in the 20th century
Uses of zinc
Zinc – for the roof over your head
Why do metal roofs need replacing?
The different ways of coating with zinc

Part 2 practical work (pre-16)

Teacher's notes..18
Worksheets
Using carbon to extract copper from copper oxide
Zinc to the rescue
Zinc in cells and batteries

Part 3 Post-16

Teacher's notes 22
Answers..22
Worksheet
Comprehension exercise *The thermodynamics of zinc extraction*

ZIRCONIA

Introduction 27
Teacher's notes..28
Zirconium and zirconia 28
Uses of zirconium and its compounds 31
Zirconium on the Internet 33

Student material
Answers..34
Worksheets
Zirconium and its compounds
Questions

Introduction

Although zinc and its extraction are not directly included in pre-16 and post-16 syllabuses, the processes by which it is isolated and its chemistry can be used to illustrate many aspects of chemistry in these courses. This booklet is designed to provide stimulating and up-to-date material for both pre-16 and post-16 courses.

Acknowledgements

The Royal Society of Chemistry thanks Britannia Zinc and in particular their Projects Manager, Bev Harris, who gave freely of his time and expertise both during the workshop and afterwards. It also thanks Rio Tinto plc for its generous sponsorship of the publication and The Institute of Materials and The Worshipful Company of Armourers and Brasiers for their support of the workshop. Catherine Gater of the Science Museum and Alan Bryant of the Kingswood Heritage Museum Trust helped to research pictures of early zinc-making processes.

The teachers were:

Martin Clarke, Newcastle Royal Grammar School, Newcastle-upon-Tyne;
Peter Ellis, St Mary's School, Wantage;
Trefor Jones, Maidenhill School, Stonehouse, Gloucestershire;
Rodney Priest, Dauntsey's School, Devizes; and
David Woolcock, St Austell College, St Austell

The structure of the material

The material is in three parts.

Part 1 (pre-16) starts with an information sheet *Zinc at Avonmouth* which describes the extraction of zinc by smelting. This is used as a basis for the comprehension exercises *Zinc smelting at Avonmouth* and *The extraction of zinc* (and also as background for the post-16 exercise in Part 3). It is followed by several stand-alone worksheets for pre-16 students. These worksheets can also be used for extension work, for homework or for work during teacher absence. Answers are provided (for the teacher) to all the questions on the sheets.

Part 2 contains worksheets for traditional pre-16 chemistry practical work, which is given a different slant by focusing on zinc.

Part 3 is a worksheet for post-16 students which concentrates on the thermodynamics of zinc extraction rather than the more usual extraction of iron. It is based on the information sheet *Zinc at Avonmouth* in Part 1.

Further information and resources
J. Emsley, *Zinc (Zn)* (Hobsons Science Support) Cambridge: Hobsons, 1992
This covers zinc production, the uses of zinc and some of the chemistry of zinc. It has questions and is suitable for post-16 students.

E.J. Cocks and B. Walters, *A history of the zinc smelting industry in Britain*. London: Harrap, 1968.
This is a comprehensive history, though now somewhat dated.

Further publications about zinc and its applications are available from

Zinc Development Association
42 Weymouth Street
London
W1N 3LQ
Tel: 0207 499 6636

The Kingswood Heritage Museum holds a collection of material relating to the life of William Champion. For further details or to arrange a visit, contact the museum at Tower Lane, Warmley, South Gloucestershire.

More general information on the extraction of other metals – aluminium, copper and iron – is available in the Royal Society of Chemistry video *Industrial Chemistry for Schools and Colleges* and its supporting materials. Details are available from the Society at the Burlington House address.

Part 1 Pre-16

RS•C

Teacher's notes
Zinc extraction at Avonmouth

This section describes the extraction of zinc, and can be used with two different sets of questions to test comprehension. Further stand-alone worksheets cover other aspects of the chemistry of zinc. This part contains:

- The information sheet *Zinc at Avonmouth*. This covers how zinc is extracted at Avonmouth by smelting. Three of the figures (sintering, smelting and refining) for this information sheet are printed separately so that the sheet can be supplied to pupils with or without them.

  Zinc at Avonmouth is designed for students to read, but it can also be used for teacher background information. It forms the basis of the following comprehension exercises:

- *Zinc smelting at Avonmouth* – a comprehension exercise based on the information sheet (using all the Figures).

- *The extraction of zinc*. This is a comprehension exercise based on the same information sheet but without using the figures sintering, smelting and refining. Instead, versions of these figures which are only partially labelled are supplied to the students with the questions, and they are asked to complete the labelling as part of the exercise. This exercise is easier than *Zinc smelting at Avonmouth*.

- *Zinc – ancient mystery, modern marvel* – a stand alone sheet suitable for classwork or homework for 11–16 year olds.

- *'Tutty' – the metal that turns copper into 'gold'* – a stand alone sheet suitable for classwork or homework for 14–16 year olds. Pupil access to data on the melting points of metals, either in book or database form, is required.

- *In the beginning, there was….ZINC* – a stand alone sheet suitable for classwork or homework for 14–16 year olds.

- *William Champion – Zinc smelter* – a stand alone sheet suitable for classwork or homework for 14–16 year olds.

- *From horizontal to vertical – Zinc smelting in the 20th century* – a stand alone sheet suitable for classwork or homework for 14–16 year olds.

- *Uses of zinc* – a stand-alone sheet suitable for class or homework for 14–16 year olds (It is useful to have *The different ways of coating with zinc* available as a reference to help answer question 4).

- *Zinc – for the roof over your head* – a stand alone sheet suitable for classwork for 14–16 year olds. This needs samples of the ores malachite, galena and sphalerite plus samples of the oxides and carbonates of zinc, lead and copper or student access to data with information about the colours of these substances.

- *Why do metal roofs need replacing?* – a follow on sheet for Zinc – for the roof over your head

- *The different ways of coating with zinc* – a stand alone sheet suitable for classwork or homework for 11–16 year olds.

Zinc

RS•C

Answers

Zinc smelting at Avonmouth

1. **a)** It is near to Avonmouth docks where the raw materials are imported by ship

   b) **(i)** zinc sulfide + oxygen → zinc oxide + sulfur dioxide
   $$2ZnS + 3O_2 \rightarrow 2ZnO + 2SO_2$$

   (ii) zinc oxide + carbon monoxide → zinc + carbon dioxide
   $$ZnO + CO \rightarrow Zn + CO_2$$
   or zinc oxide + carbon → zinc + carbon monoxide
   $$ZnO + C \rightarrow Zn + CO$$

   (iii) Reduction here means loss of oxygen

   c) A temperature > 1000 °C ensures that zinc is produced as vapour. If the temperature < 1000 °C,
   $$Zn + CO_2 \rightarrow ZnO + CO$$
   takes place rather than the reverse reaction.

   d) Quenching is rapid cooling of Zn(g) to Zn(l). It minimises the reaction
   $$2Zn(g) + O_2(g) \rightarrow 2ZnO(s).$$

2. **a)** **(i)** $2SO_2 + O_2 \rightarrow 2SO_3$

   (ii) $SO_3 + H_2O \rightarrow H_2SO_4$

   b) 98 tonnes

   c) 92.7 tonnes

3. **a)** **(i)** Lead is associated with potential brain damage and is a cumulative poison.
   Pregnant women could risk harm to their babies.
   Traditionally there are fewer women employed in industry.
   Sex discrimination may have operated.

   (ii) Uses of lead include: lead-acid (storage) batteries; lead crystal glass; pigments in chemicals; a shield from radiation (eg X- or gamma-rays); building/roofing material.

   b) Carbon monoxide – a poisonous gas
   Sulfur dioxide – can cause breathing problems, produces acid rain

   c) Slag has to be disposed of safely. This is particularly costly since the introduction of the landfill tax.

4. **a)** Zinc ores are bought and sold in dollars.

   b) Some reasons for the fall in work force might be –
   greater automation, computerisation, need to reduce costs, improved efficiency.

   c) *Community Link* provides good public relations, keeps the local community informed, keeps the local community involved, maintains (or establishes) good relations, counteracts prejudice or misinformation.

d) The correct mix is chosen to give an economic yield and depends on: the required composition of product, the availability of concentrate, the cost of concentrate.

5 a) Transition metals

b) The density of liquid zinc is less than that of liquid lead. Zinc floats on lead.

c) (i) Cadmium has a lower boiling temperature than zinc. Cadmium boils off first, or cadmium boils off from the top of the tower.

(ii) Fractional distillation

(iii) Other industrial separations are – separation of air, separation of crude oil.

Extraction of zinc

1. Figures for 'sintering', 'smelting' and 'refining' with labels filled in.

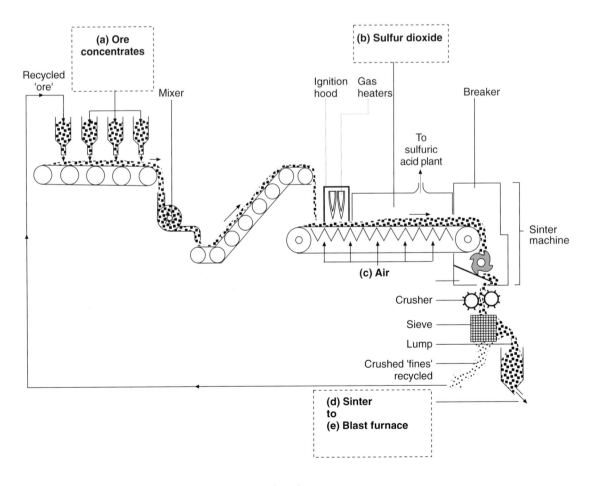

Sintering

Zinc

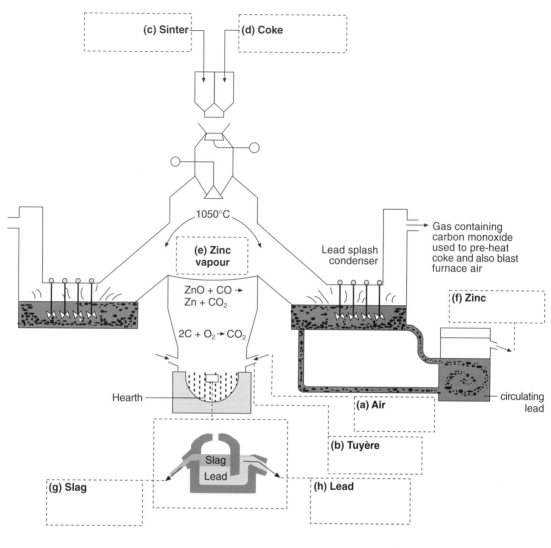

Smelting

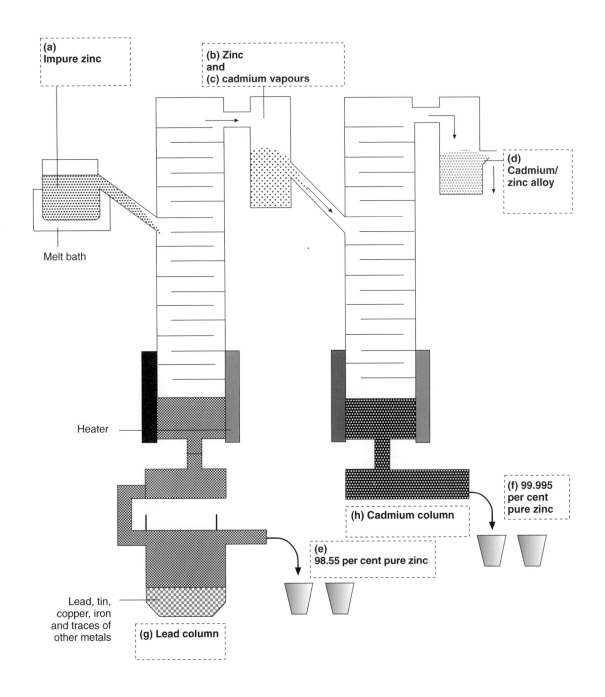

Refining

Zinc

b) A large surface area allows more gas/solid reaction to take place with the carbon monoxide gas.

c) Zinc oxide loses oxygen to become zinc. This is a reduction.

d) Zinc oxide + carbon monoxide → zinc + carbon dioxide
   $ZnO + CO → Zn + CO_2$

e) Zinc has a much lower boiling temperature than iron and, unlike iron, can be boiled off as it is formed.

2. a)

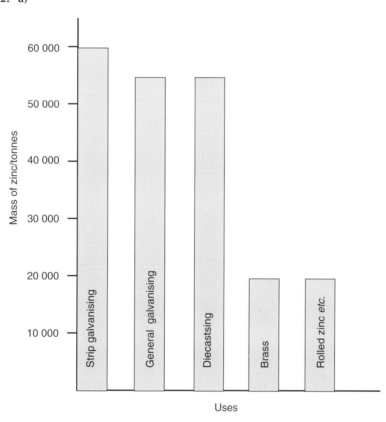

b) i) Pie chart – this gives the per cent distribution of how zinc is used.

ii) Block graph – this gives the absolute numbers in tonnes.

Zinc – ancient mystery, modern marvel

1. a) Suitable properties might include: waterproof, weather resistant, tough, fire resistant, resistant to chemicals, available in small units, good appearance, easy to replace/repair, not too dense.

2. Zinc metal had only just been isolated and it must have been rather an unknown quantity.

3. Galvanised steel has a thin coating of zinc. This protects the steel from rusting even when the coating is damaged.

4. Uses of brass include:
   taps, door knobs, ships' instruments   corrosion resistant
   ornaments,   attractive appearance,
   electrical connections,   good conductor of electricity
   eg electrical plug pins.

'Tutty' – the metal that turns copper into 'gold'

1. Copper, silver, gold, tin, lead, iron, mercury

2. zinc + oxygen → zinc oxide
   $2Zn + O_2 \rightarrow 2ZnO$

3.
Metal	Melting point/°C	Boiling point/°C
Copper	1083	2567
Lead	328	1740
Silver	962	2212
Zinc	420	907

   Zinc has a lower boiling temperature than the others.

4. Carbon + zinc oxide → carbon monoxide + zinc
   $C + ZnO \rightarrow CO + Zn$
   or Carbon + zinc oxide → carbon dioxide + zinc
   $C + 2ZnO \rightarrow CO_2 + 2Zn$

5. Uses of zinc might include – making die-cast models and components, rust prevention, making brass alloys, making zinc oxide (used in tyres and medicinal creams), batteries, zinc sheeting, galvanising.

In the beginning there was ZINC

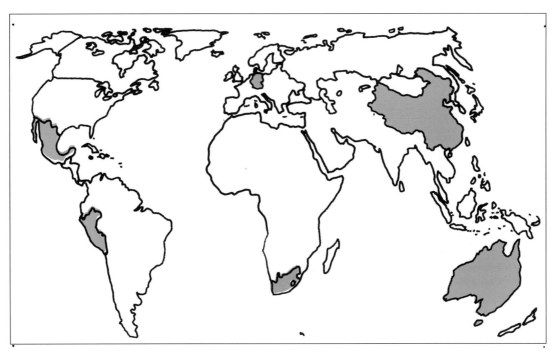

1. Map

2. The stacked paper would be roughly 100 m tall (depending on the paper thickness).

3. a) Sedimentary rock is made of layers of particles of rock laid down over millions of years.

Zinc

RS•C

b) Igneous rock is formed from cooled magma.

c) Metamorphic rock is rock that has undergone rearrangement of its structure by the action of high pressures and temperatures.

d) An igneous intrusion is a rivulet of magma that has cooled to a solid under the ground, without reaching the surface.

4. The crystals will be large. The liquid magma will have cooled slowly under the surface, and slow cooling produces large crystals.

5. An ore is a rock which has enough mineral in it to make it worth mining commercially.

6. a) The iron ball bangs against the side of the vessel crushing any lumps to powder.

b) The tiny fragments of metal ore cling to the surface of the bubbles and come to the top on the floating froth.

7. In the furnace the zinc oxide is losing oxygen leaving zinc on its own. (This is a reduction process).

William Champion – zinc smelter

1. Essentially paragraph 5.

2. The zinc vapour cooled to a liquid.

3. Any oxygen reacts with the zinc to give zinc oxide.

4. a) Zinc oxide + carbon → zinc + carbon monoxide
   $ZnO + C \rightarrow Zn + CO$

   b) Reacting ratios:
   | 81 | 12 | 65 | 28 |
   | 1.62 kg | 0.24 kg | 1.3 kg | 0.56 kg |

   ie 1.62 kg of zinc and 0.24 kg of carbon.

   c) Zinc oxide 1.62 x 8 = 12.96 kg, carbon 0.24 x 8 = 1.92 kg

   d) More needed because the process is neither 100 per cent efficient, nor the ore and charcoal pure.

5. Mark by impression to evaluate communication skills.

From horizontal to vertical – zinc smelting in the 20th century

1. Easier automation; furnace does not have to cool between batches; no time lost when retorts have to be emptied of slag.

2. Less fuel used, lower labour costs.

3. $ZnO + CO \rightarrow Zn + CO_2$

4. $Zn + CO_2 \rightarrow ZnO + CO$

5. Capital costs of building new plant are too high in some countries. If labour costs are very low, labour-intensive processes are favoured (despite health and safety considerations).

6. U-boat attacks on convoys from the US could have left the UK without the zinc necessary to make arms to fight. Bullet and shell cases are made of brass – an alloy of zinc and copper.

7. To prevent the UK from depending on foreign supplies in a future conflict; to maintain employment.

8. Avonmouth has a deep water port through which zinc ores can be imported (there are no commercially sized deposits of zinc in the UK) and good rail links (and now roads). It is near the South Wales coal field (which used to be the main source of coke), and near a source of labour (Bristol).

Uses of zinc

A Zinc for die-casting metal toys
1. Lead is toxic.
2. Plastic, because the paint solvent affects the plastic.
3. The lower the temperature, the cheaper and safer the process.
4. Steel and hot, molten zinc might form an alloy.

B Brasses
5. Brass is a harder and stronger metal alloy than copper and will therefore resist wear on being plugged and unplugged. Copper wire is flexible and cheaper than brass.
6. Any suitable suggestions of uses of brass, such as plumbing components or door handles. Comments on suitability depend on the use suggested.

C Zinc for making zinc oxide
7. Properties for a tyre material might include - elasticity, resistance to wear, to chemicals and to change at high temperature.
8. Zinc oxide prevents the growth of bacteria and fungi and is non-toxic.

Zinc – for the roof over your head.

1. Zinc, lead, copper.
2. Sphalerite – white
   galena – black
   malachite – dark green.
3. Zinc carbonate – white
   lead carbonate – white
   copper carbonate – green
   zinc oxide – white
   lead(II) oxide – white
   copper(II) oxide – black.
4. The oxide and the carbonate of the metal are coating the metal. In the case of copper, first the oxide and then the carbonate form a coating.
5. a) zinc + oxygen → zinc oxide
      $2Zn + O_2 \rightarrow 2ZnO$

   b) copper oxide + carbon dioxide → copper carbonate
      $CuO + CO_2 \rightarrow CuCO_3$

Zinc

Why do metal roofs need replacing?

1. The roofing metal becomes thinner and eventually cracks.

2. In climates with extremes of temperature.

3. a) zinc oxide + sulfuric acid → zinc sulfate + water
   $ZnO(s) + H_2SO_4(aq) \rightarrow ZnSO_4(aq) + H_2O(l)$

   b) zinc hydroxide + sulfuric acid → zinc sulfate + water
   $Zn(OH)_2(s) + H_2SO_4(aq) \rightarrow ZnSO_4(aq) + 2H_2O(l)$

   c) zinc carbonate + sulfuric acid → zinc sulfate + carbon dioxide + water
   $ZnCO_3(s) + H_2SO_4(aq) \rightarrow ZnSO_4(aq) + CO_2(g) + H_2O(l)$

4. a) zinc oxide + nitric acid → zinc nitrate + water
   $ZnO(s) + 2HNO_3(aq) \rightarrow Zn(NO_3)_2(aq) + H_2O(l)$

   b) zinc hydroxide + nitric acid → zinc nitrate + water
   $Zn(OH)_2(s) + 2HNO_3(aq) \rightarrow Zn(NO_3)_2(aq) + 2H_2O(l)$

   c) zinc carbonate + nitric acid → zinc nitrate + carbon dioxide + water
   $ZnCO_3(s) + 2HNO_3(aq) \rightarrow Zn(NO_3)_2(aq) + CO_2(g) + H_2O(l)$

The different ways of coating steel with zinc

1. Hot dip galvanising – produces hard, long-lasting coat.
   (spraying is the next best).

2. Sherardising – it can cope with complicated shapes and it is cheap.

3. Spraying will reach all the parts of the bridge and give it a good thickness of coat. It can be renewed whenever it is needed.

4. Zinc dust painting – it is suitable for small localised jobs.

Zinc at Avonmouth

The properties of Zinc (Zn)

- Atomic number　　　　30
- Relative atomic mass　65.4
- Melting temperature　420 °C
- Boiling temperature　907 °C
- Density　　　　　　　7.1 g/cm^3

Zinc is a very important metal. It is the fourth most widely used in the world. Seven million tonnes of zinc metal is refined every year worldwide. Its principal uses are shown in the pie chart. Although it has been used for well over 2 000 years, especially in the form of the alloy brass (brass is copper with up to 40 per cent zinc), pure zinc was not isolated until about 200 years ago.

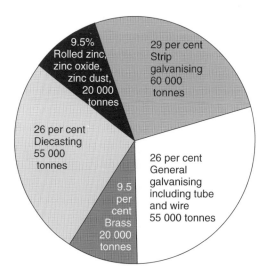

UK zinc market by end use – 1995 figures, 210 000 tonnes total

Zinc may be isolated from its ores by smelting (heating with carbon), in much the same way as iron, or by electrolysis. About 20 per cent of the world's zinc metal is produced by smelting. Unlike electrolysis, the smelting process can cope with a varied composition of ore input. Britannia Zinc at Avonmouth near Bristol, the only manufacturer of zinc in the UK, uses a blast furnace which smelts both lead and zinc at the same time.

Zinc and lead smelting at Avonmouth

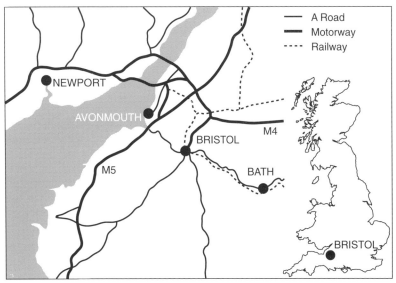

Map of Avonmouth

The zinc extraction process

Ores

The most important zinc ores are shown in the table

Zinc ores

Name	Chemical name of main zinc mineral	Formula
Zinc blende	Zinc sulfide	ZnS
Marmatite	Zinc iron sulfide	$ZnFeS_2$
Calamine	Zinc carbonate	$ZnCO_3$

Shipment of ore concentrates

Approximately 320 000 tonnes each year of ore concentrates, (ores which have been ground and treated to raise the zinc content to 50 per cent or more) are imported into Avonmouth Docks from Australia, South America, Canada, Alaska, Iran, Scandinavia and Southern Ireland. A conveyor belt, approximately 1 km in length, is used to carry up to 600 tonnes/hour from the docks to the works. Here, the ore is weighed automatically at the conveyers before being deposited in a store on site which holds up to 50 000 tonnes under cover. From start to finish all the processes are regulated and monitored by computers and instruments in the centralised control room.

Sintering

The materials used in the zinc smelting blast furnace have to be strong enough to support the load above them in the furnace, must include zinc and lead as oxides, and must be porous and have a large surface area. This is done by producing sinter, a pumice-like material which is like a hard sponge in structure.

First the ore concentrates from various sources are mixed so that they will give the required yields of zinc, lead and other metals. Recycled materials are also used.

Then the mix is fed onto the sinter machine. This has iron bars – larger than, but similar to, those which form the base of a domestic coal fire – which are part of a moving trolley. Air passes up through the bars and the sinter mix is ignited using natural gas burners. The material fuses into lumps called agglomerate. It becomes porous as, during the process, sulfur is burnt off as sulphur dioxide gas. This gas is collected and converted into sulfuric acid by the Contact Process in which the sulfur dioxide reacts with further air and with water. The metals are now present as oxides.

The sintering process is shown in the figure 'Sintering'.

Smelting

The blast furnace at Avonmouth (the imperial smelting furnace or ISF) uses coke in very large quantities (up to 130 000 tonnes per year) provided by the Cwm coke ovens in South Wales. Every tonne of zinc produced requires nearly 0.9 tonnes of coke.

Hot air at 1000 °C is blown in at the base (hearth) of the furnace, through pipes called tuyères, and coke and sinter are added at the top.

The coke burns in the air to produce carbon dioxide, carbon monoxide and intense heat. In the main process, zinc oxide is reduced to zinc by carbon monoxide (and carbon). The temperature at the top of the furnace is kept above 1000 °C for two reasons – to ensure that zinc is present as a vapour (it boils at 907 °C) and because at lower temperatures zinc reacts with carbon monoxide to go back to zinc oxide and carbon.

At the same time, lead oxide, which is also present in many of the ores, is reduced to lead and falls to the bottom of the furnace into the hearth.

The smelting process is shown in the figure 'smelting'.

Separation

Since zinc is produced as a vapour, it passes out in the gas stream from the top of the furnace. The molten lead (with other metals such as silver, gold and copper) and slag, a material formed from impurities, sink to the bottom of the furnace. They are tapped off through water-cooled copper blocks into the forehearth.

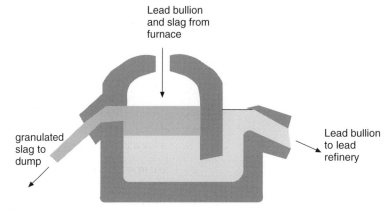

The forehearth – slag is separated from denser lead bullion

The slag and lead tapping is a spectacular process which is carried out around every 1½ hours. The operator, wearing protective clothing, uses an oxygen lance to melt a hole in the solid slag plug in the copper block. The bright, molten river of slag and lead bursts out with a flying, sparkling spray. After 20 minutes or so a water-cooled check bar is inserted into the tap hole to cool the slag/lead melt in the hole and re-seal the hole. The lead separates to the bottom of the forehearth and flows out into moulds; the slag is run out from the top. The lead is recovered together with its silver, gold and copper content as four-tonne blocks. Granulated slag is a fairly inert material and is transported to a suitable dump. Research to find uses for the slag is currently in progress. It is formed from impurities in the ore concentrates such as lime (calcium oxide), aluminium oxide and sand (silica).

Slagging the furnace

Condensing the zinc

A lead splash condenser is an important feature of the Avonmouth process. This process rapidly cools (quenches) the zinc vapour from 1000 °C to 550 °C.

- Zinc vapour from the furnace passes into the condenser.

- In the condenser, the zinc vapour dissolves in a spray of fine droplets of lead, produced by rotors splashing a pool of molten lead.

- Zinc dissolved in molten lead is cooled further. It flows into a separator bath where the zinc floats on top of the molten lead, is separated and then taken to the refinery.

Refining

The zinc from the imperial smelting furnace (98.5 per cent purity; but still containing some lead, cadmium, tin, copper, iron and other metals) is refined in a two stage distillation process.

- The zinc is fed into a 'lead' column where approximately 33 per cent of the zinc is boiled off from the top, together with all the cadmium.

- This mixture is then cooled to a liquid and fed to a 'cadmium' column to produce a cadmium alloy and a special high grade zinc (up to 99.995 per cent purity).

- The liquid zinc from the base of the 'lead' column is separated before being treated with metallic sodium to remove small traces of arsenic. This cast zinc (98.55 per cent pure) is suitable for galvanising and making brass.

The process is very similar to the fractional distillation of crude oil – the metals with the lower boiling points come off at the top of the column, while the metals with the highest boiling points remain at the bottom.

The refining process is shown in the figure 'Refining'.

Economics

Zinc production is an international operation. The ore concentrates and other raw materials are bought and the products sold in US dollars. This means that changes in the $/£ exchange rate can significantly affect the profitability of the company.

Environment

The authorised limit of lead emissions from the whole plant is 3.0 kg per hour. The company gives a quarterly magazine *Community Link*, to local residents, which informs them of current emission levels. Since 1989/90 emissions of lead have been kept well below the permitted level through continuous process improvement.

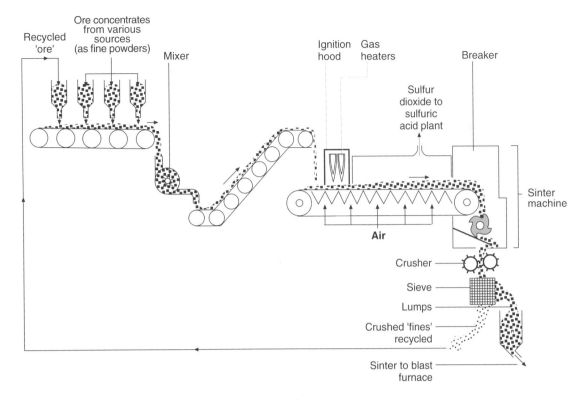

Sintering

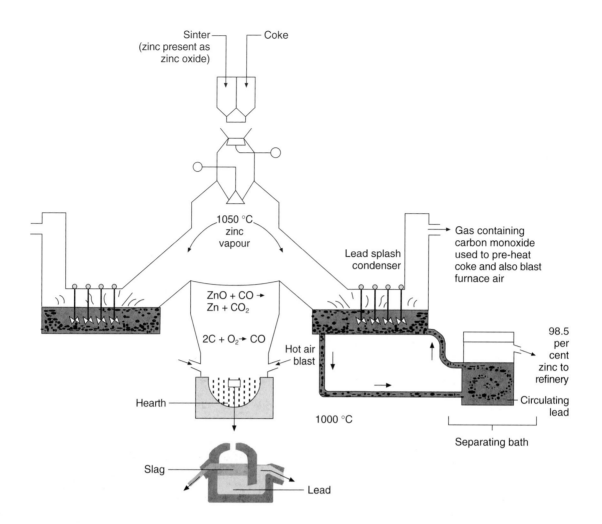

Smelting

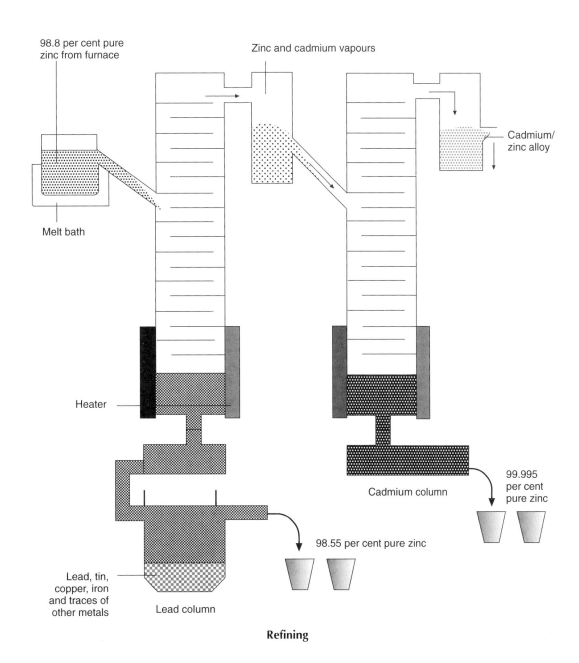

Refining

Zinc smelting at Avonmouth

Use the passage and your knowledge of chemistry to answer the following questions.

1. (a) Why is Avonmouth a suitable location for the zinc smelting works?

   (b) Write word and then balanced chemical equations to represent:

   (i) zinc sulfide (ZnS) being heated in air (oxygen) to give zinc oxide and sulfur dioxide;

   (ii) zinc oxide (ZnO) being reduced to zinc in the furnace by either carbon monoxide, CO, or carbon, C.

   (iii) Explain simply what is meant in the passage by the term *reduced*.

   (c) Give two reasons why the furnace temperature has to be carefully controlled.

   (d) The passage refers to the *rapid quenching* of zinc vapour. Explain what this means and why it is important in this process.

2. In the zinc smelting process, sulfur dioxide is produced which is made into sulfuric acid by the Contact Process.

   (a) Write balanced chemical equations for the following reactions which are part of the Contact Process:

   (i) sulfur dioxide reacting with oxygen to give sulfur trioxide

   (ii) the reaction of sulfur trioxide with water to give sulfuric acid.

(b) Calculate the maximum mass of sulfuric acid which could be produced from 64 tonnes of sulfur dioxide.
(Relative atomic masses: H = 1; S = 32; O = 16.)

(c) If the sulfur dioxide came from pure marmatite, how much of this mineral would be needed?

3. (a) The lead industry is always under close scrutiny as far as health and safety issues are concerned. Employees undergo regular and frequent blood tests. Until very recently women were not allowed to work on the smelting plant.

　(i) Suggest reasons for these statements.

　(ii) Give one important use of lead.

(b) Wherever possible in this works, materials and chemical plant are under fume hoods so that the vapours can be removed and treated.

　(i) Give two harmful vapours (apart from lead) present in this smelting works and say how each is potentially harmful.

(c) Suggest one problem posed by the slag

4. (a) Suggest why managers in the zinc works keep a close watch on the US $/£ sterling currency exchange rates.

(b) The number of employees at the plant has fallen from 750 to around 500 in recent years. Suggest two reasons for this.

(c) Write down two reasons a manager in the company might give for publishing *Community Link* and sending it to local residents.

_____
_____

(d) Suggest two factors which would influence the proportion of ores that are mixed to make the furnace feed.

_____
_____

5. a) The lead bullion produced contains copper, silver and gold. What name is given to the area of the Periodic Table containing copper, silver and gold?

_____

(b) Deduce from the passage which is denser, molten lead or molten zinc. Give a reason to support your answer.

_____
_____

(c) The difference in boiling points of zinc and cadmium is used to separate them.
  (i) Deduce from the passage which has the higher boiling point, zinc or cadmium. Give a reason to support your answer.

  _____

  (ii) What name is given to this type of separation process?

  _____

  (iii) Give one other important industrial application of this separation process.

  _____
  _____

Zinc smelting at Avonmouth: page 3 of 3

Zinc extraction

1. a) Use the passage *Zinc at Avonmouth* to help you fill in the labels on the diagrams 'sintering', 'smelting' and 'refining'.
   Write short notes under each diagram to explain what is happening.

   b) Suggest why a large surface area is useful for the material for the blast furnace produced after sintering.

   _____
   _____
   _____

   c) Explain why the process of obtaining zinc from zinc oxide in the blast furnace is called a reduction.

   _____

   d) Write a word and balanced symbol equation to represent this reduction, which uses carbon monoxide (CO) as the reducing agent and produces carbon dioxide (CO_2) in the process.

   _____
   _____

   e) Iron is also extracted in a blast furnace, but it is not separated from impurities in the same way as zinc. What difference in the properties of the two metals explains this.

   _____
   _____

2. a) Represent the pie chart on the 'uses of zinc' as a block graph (using the values for tonnes)

   b) Which way of presenting data would be most useful for:

      i) a zinc manufacturer who must decide in which form to produce zinc

      _____
      _____

      ii) a zinc manufacturer who must decide how much ore to import? Explain your answers.

      _____
      _____

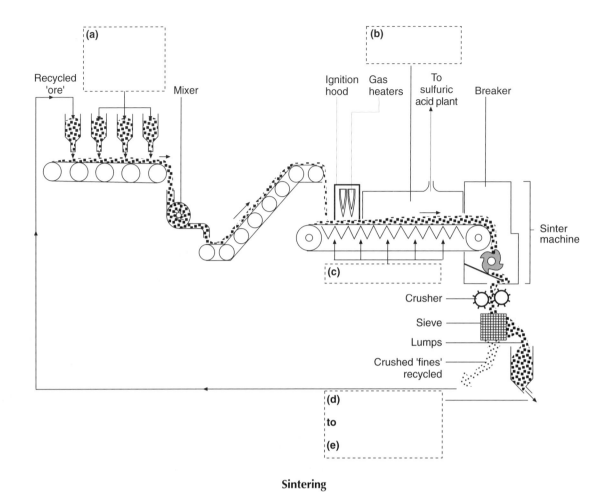

Sintering

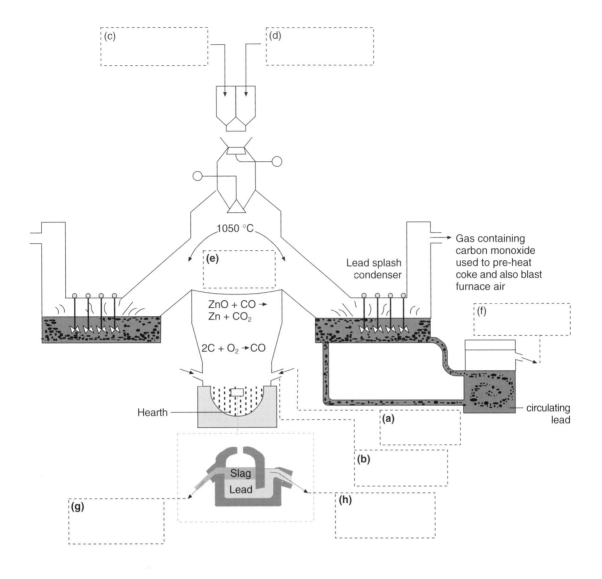

Smelting

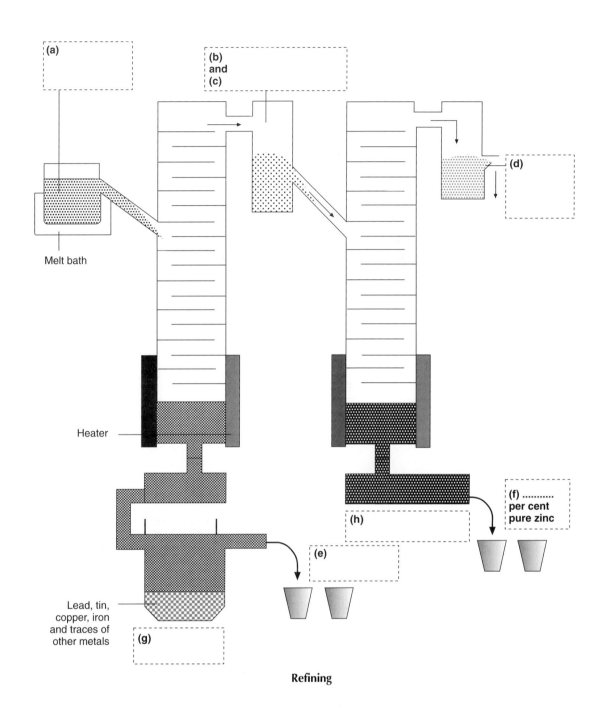

Melt bath

Heater

Lead, tin, copper, iron and traces of other metals

Refining

Zinc – ancient mystery, modern marvel

Blue roofs

The roofs of Paris

When the architect Baron Haussman was rebuilding Paris after the Napoleonic wars of the early nineteenth century he had a vision of a new city with wide streets and elegant buildings. He chose zinc sheet as the material to roof all the buildings. Zinc does not corrode, but when it weathers it produces a light blue effect which is now one of the features of France's capital city. Some zinc roofs have lasted over 100 years without having to be repaired.

Baron Haussman's choice was a bold one, because it was only in 1805 that a way of rolling zinc sheet was invented and in the early 1800s the extraction of zinc from its ores on a large scale was very new.

In 1836 another use for zinc was found. This was galvanising. Carefully prepared steel sheet is dipped in molten zinc, giving the steel a thin, protective coating of zinc. The layer of zinc does not rust and only corrodes very slowly, developing a layer of zinc oxide and zinc carbonate. Even when scratched the remaining zinc continues to protect the iron. Corrugated galvanised iron sheet soon became a cheaper if less elegant form of lightweight roofing material than zinc.

New England galvanised roofs

While the benefits of galvanising iron were recognised, one of the main uses of zinc up to the 20th century was in making an alloy that had been used for hundreds of years – brass.

Questions

The table below compares different materials used to make roofs.

1. Fill in the table with the answers to the following.

    a) Make a list of the properties needed for a material to be a good choice for a roof and put them in the column marked 'properties'.

    b) Go down the list and tick if a thatched roof has this property.

    c) Choose another material that you know is used for roofs. Write its name in the space in the table and tick the list if it has the property.

    d) Now do the same for zinc as a roofing material.

Properties	Thatch	Other material	Zinc

2. Why was it a bold move to use zinc for roofs in the early 1800s?

3. Dustbins are often made from galvanised steel. What is galvanised steel, and why is it better than plain steel?

4. Zinc is used to make brass. Give examples of uses of brass and for each use explain why brass is a suitable material.

Zinc – ancient mystery, modern marvel: page 2 of 2

'Tutty' – the metal that turns copper into 'gold'

Ancient civilisations (before about 1600 BC) knew only seven metals and zinc was not one of them. However they knew that if copper was mixed with a whitish mineral, called calamine, and charcoal and then heated fiercely in a fire the copper became changed into a hard, golden metal that we call brass.

1. Suggest the names of seven metals known to the ancients? (Hint: think about the reactivity series or use a data book, database or CD-ROM.)

   _____
   _____
   _____
   _____

Calamine is impure zinc carbonate but when it is heated alone with charcoal no metal is produced. The metal smelters did not realise that zinc was being formed as a vapour and that when it met the atmosphere it immediately re-oxidised to form zinc oxide.

2. Write a word and a symbol equation for the reaction of zinc vapour with oxygen.

   _____
   _____

The metal workers of Rajastan in Northern India were the first to solve the problem of extracting zinc around about the 14th century. The Indians had long experience in smelting copper, lead and silver and also had access to large amounts of zinc ore.

3. Use a data book or CD-ROM to find out the melting and boiling points of copper, lead, silver and zinc. What do you notice about the boiling point of zinc?

   _____
   _____
   _____
   _____
   _____

After first roasting the ore to get rid of sulfur it was mixed with charcoal and a number of other materials such as treacle, which were supposed to have magical properties. The mixture was then packed into clay retorts about the size and shape of an aubergine,. A wooden stick pushed into the retort would, when burnt out, provide an escape hole for the zinc vapour.

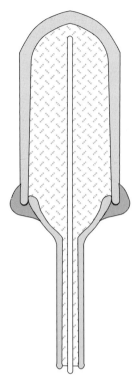

A clay retort used for extracting zinc

The retorts were stacked in a closed furnace and heated to over 1100 °C. The zinc ore was reduced to zinc which boiled off. The zinc vapour escaped from the retorts and was collected in a condenser. Because air was not allowed into the furnace, the zinc was prevented from returning to its oxide. The liquid metal was cast into ingots.

4. Charcoal is mainly the element carbon. Write a word and symbol equation to show how carbon reacts to reduce zinc oxide to zinc.

_____
_____
_____
_____

In the 17th century the Chinese copied the Indian method and samples of zinc began to arrive in western Europe. There was great excitement over this new metal which was known as 'tutty'. The method of its extraction and even the ore it was obtained from were unknown in Europe but it was very useful for mixing with copper to make better brass alloys. In the early 1700s zinc imported from the East was very expensive. However, the secrets of its origin were known by this time, and calamine, the main zinc ore, was cheap and available. It was not long before a European discovered the Indian methods.

5. Find out some of the uses of zinc.

_____
_____
_____
_____

In the beginning, there was... ZINC

Where in the world do we mine zinc?

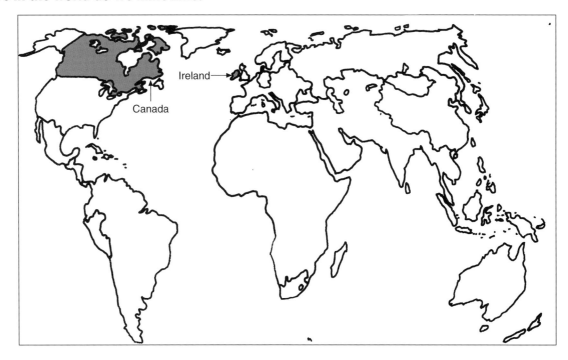

1. Use an atlas to mark on the map these other zinc producing countries: Mexico, Germany, Peru, South Africa, China, Australia.

   Is zinc abundant? Here are the average concentrations of zinc worldwide:

Rocks	70 ppm
Soil	150 ppm
Fresh water	20 ppm
Sea water	1 ppm

   (ppm = parts per million)

2. To help you to imagine what 70 ppm means, try the following. Count the number of pages in an exercise book
   Write down roughly how many books you need for a million (1 000 000) pages.

   Work out and write down how tall a stack of this many books would be.

   Now imagine hiding 70 £5 notes at random in the stack.

   Would it be easy to find the £5 notes?

   You can see that, on average, zinc is not a very abundant element (in fact it is slightly less abundant than nickel and slightly more abundant than copper). Luckily, natural processes in the rock cycle have concentrated zinc minerals in several places. When the concentration is high enough, it may be worthwhile mining this ore body.

The formation of zinc deposits

The figure below shows some of the ways that the rock cycle gathers together (concentrates) zinc minerals.

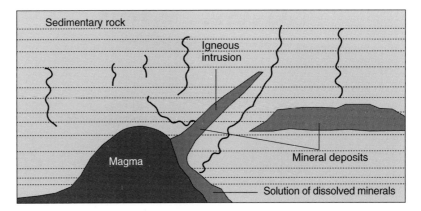

Formation of mineral deposits

- Water flows down rock cracks and becomes trapped deep within the Earth's crust.
- Compounds (called minerals) of zinc and some other metals, which are present in the magma, will dissolve in hot, pressurised water so a solution forms.
- This can be carried upwards in an igneous intrusion, or pass upwards into sedimentary rocks (often limestone).
- The minerals are then deposited as the solution cools.
- Sedimentary or igneous rocks can be metamorphosed (changed) by heat and pressure, still carrying their enriched mineral deposits. There is a a very large deposit of zinc minerals in Australia which was formed in this way.

3. What is meant by
   a) sedimentary rock

   _____

   _____

   b) igneous rock

   _____

   _____

   c) metamorphic rock?

   _____

   _____

   d) an igneous intrusion?

   _____

   _____

4. Would you expect the crystal size of the rock in an igneous intrusion be

large or small? Explain your answer.

_____
_____

A commercial ore might contain 50 000 ppm of zinc. Now you're looking for 50 000 £5 notes in the stack of pages.

5. What is meant by the word ore?

_____
_____

Even a rich ore body has lots of useless sandy minerals in it. These must be removed and the ore (such as zinc sulphide) concentrated. Nowadays it is done by crushing followed by froth flotation

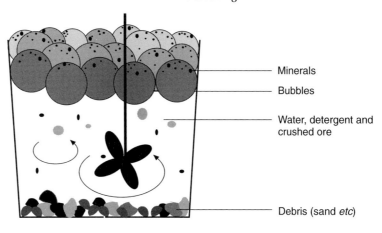

1. Crushing

2. Froth flotation

6. a) How does the crushing process work?

_____
_____

b) Why is process 2 called froth flotation?

_____

_____

After further processes the ore is ready for smelting in a blast furnace.

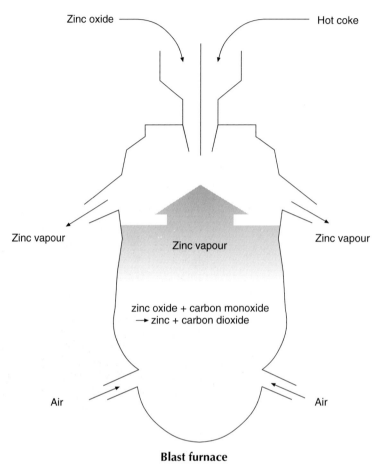

Blast furnace

7. What is happening to the zinc oxide in the furnace?

_____

_____

William Champion – zinc smelter

William Champion began the production of zinc in the UK. William was born in 1710, the youngest member of a well-off family that lived in Bristol. William's father, Nehemiah, a landowner, had started a company in 1700 to manufacture brass. William worked for his father in the foundry as a young man but in 1732, perhaps impatient and ambitious, he left to set up his own works. There are stories that he travelled in Europe dressed in rags to learn the secrets of foreign metal smelters and that he had contacts with a sailor from China who had seen zinc smelters working.

William Champion's house in Kingswood, Bristol

In 1737 William had a zinc smelting furnace operating in Bristol and he took out a patent in 1738. By 1746 he had moved to a much larger site on the city's outskirts in the village of Warmley. Here 600 workers, mainly local but some from abroad, laboured at William's copper and zinc smelters and brass foundries and 1400 people were employed by other companies turning the brass into containers, plates, and enough wire to make ten million pins a week.

William was a Quaker and showed some concern for his employees and provided them with housing at the Warmley works. He also built a large house for himself in the fashionable Dutch style. The factory site was powered by water-wheels fed by a large artificial lake which had a huge statue of the Roman god Neptune at its centre.

William's method of smelting zinc became known as the English method. An octagonal cone-shaped furnace contained up to eight crucibles.

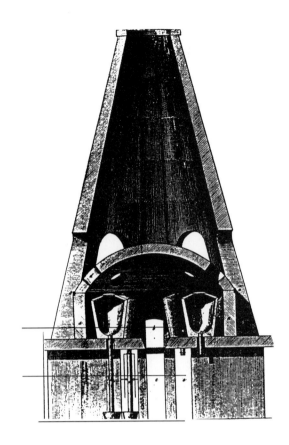

Crucible

Iron pipe

Pan of water

William Champion's furnace

After the charge of preheated calamine (zinc oxide) and charcoal (carbon) was poured into the crucible, the cover was sealed. An iron pipe passed through the bottom of the crucible and into pans of water outside the furnace. When the furnace was fired, the temperature in the crucibles rose to over 1100 °C and zinc vapour was formed which passed down the iron pipes. Zinc condensed and ran into the pans of water. No air was allowed to enter the pipe or crucible so the hot zinc was prevented from reacting with oxygen. Only about 1–2 kg of zinc was produced by each charge of the crucibles. It was very hot work for the workers looking after the fires and emptying out the slag from the crucibles.

Brass making was profitable but William's ambitions lead him into some unsuccessful business projects at Bristol docks. In 1769 he was bankrupt and forced to sell his company to his rivals in brass manufacturing, the Bristol Brass Wire Company. William died in 1789 but the works that he built continued to operate for another 100 years.

Questions

1. Highlight or underline the section in the passage about the process of making zinc.

2. Explain what is meant by 'zinc condensed and ran into the water'.

3. Why was it important that no oxygen was allowed to react with the zinc?

4 a) Write a word and a balanced symbol equation for the production of zinc from zinc oxide, assuming that carbon monoxide, CO, is produced in the reaction.

_____

_____

b) If one crucible produced 1.3 kg of zinc, how much
  i) zinc oxide and ii) carbon are needed in theory in the crucible to produce this amount?
  (Relative atomic masses: Zn = 65, O = 16, C = 12.)

_____

_____

c) How much of each is needed for the whole furnace?

_____

d) Would you expect the crucibles to need more or less than this amount, or this amount exactly? Explain your answer.

_____

_____

5. Imagine that you are one of Champion's workers employed to operate the zinc furnaces. Describe a day in your life.

From horizontal to vertical – zinc smelting in the 20th century

In 1914 nearly all the UK's zinc supplies were imported from Belgium and Germany. These two countries took almost all the zinc ore mined from the vast deposits in Australia.

The outbreak of World War I exposed the UK industry's shortsightedness. Zinc was in demand for the arms industry and for most of the war the UK relied on supplies imported from the US. The government decided that the country must not be dependent on foreign sources of zinc. After a mismanaged attempt in 1916 which foundered in 1923 the UK zinc industry was finally established in 1929 under the name National Smelting Company and later the Imperial Smelting Corporation Limited. Avonmouth, near Bristol on the Bristol Channel, had been chosen as the site back in 1916 and the first plant to be built used the traditional **Belgian method**

The Belgian process (see following page), consisted of furnaces packed with small horizontal retorts. The retorts were charged with zinc ore and coke and then sealed to stop air entering. On heating, the zinc formed as a vapour which was collected at the end of the retorts and condensed. The Avonmouth plant was intended to have 24 furnaces with 9216 retorts in all, producing 70 000 tonnes of zinc a year. It was never completed. However, the ore roasting plant was built as it gave off sulfur dioxide which could be turned into sulfuric acid, a valuable commodity.

In the 1930s Imperial Smelting adopted the **vertical retort process**, developed in the US, in which briquettes of zinc ore and coke were fed in a vertical column.

When the column was heated, the zinc boiled off to be collected in an airtight condenser. The vertical retorts were a great improvement as they required less back-breaking labour and could produce far more zinc in a continuous process. Their biggest disadvantage was the cost of fuel required to heat the outside of the retorts and the high cost of the refractory bricks, made from silicon carbide, which were needed to line the furnace.

Imperial Smelting hoped to use a blast furnace method where hot air would ignite the coke and reduce the zinc ore to zinc vapour. But blast furnaces had a big problem. The air that was blown in oxidised carbon monoxide to carbon dioxide. When the temperature of the zinc vapour and carbon dioxide fell below 1000 °C they would react to re-form zinc oxide. In 1943 L.J.Derham had an idea. He suggested that to remove the zinc, the zinc vapour from a blast furnace could be passed into a chamber containing molten lead. A paddle wheel splashed the lead, and the zinc vapour dissolved in the lead droplets. The solution of zinc in lead was then pumped off. When the mixture was cooled to approximately 430 °C, molten zinc settled out on the surface of the lead to be skimmed off. The blast furnace also produced lead from the mixture of ores that were used as the charge and cadmium, a common impurity in zinc ores, could also be recovered by redistilling the zinc.

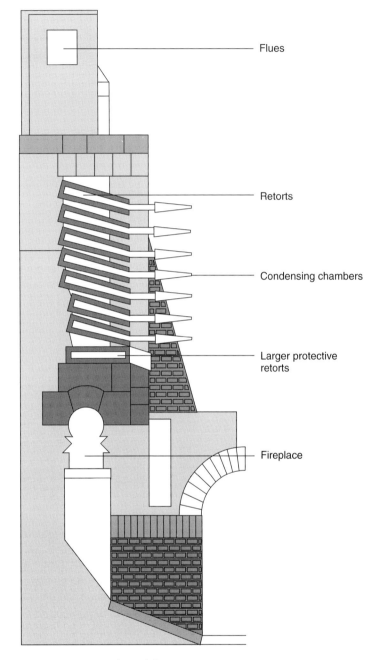

The Belgian process
(from W.R. Ingall. Lead and zinc in the United States.
Hill Publishing Company, 1808)

Experiments showed that Derham's idea worked, but there were many problems to be solved. In 1949 the first full scale Imperial Smelting Furnace was built at Avonmouth followed in 1952 by the second. Over the next 40 years the process was adopted in the many parts of the world where electrolysis of zinc ores is not an economic possibility. Also, mixed ore concentrates of zinc and lead can be used in the blast furnace, whereas the rival electrolytic process requires clean zinc concentrates. Today Avonmouth produces about 100 000 tonnes of zinc a year from one furnace and supplies nearly half of the UK's needs.

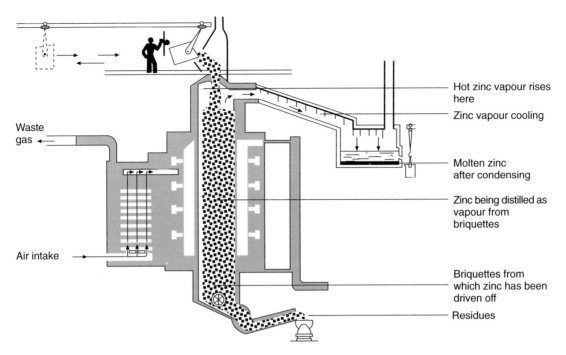

The vertical retort process
From E.J. Cocks and B. Walters,
A history of the zinc smelting industry in Britain. London: Harrup, 1968

Questions

1. The horizontal retort method was a batch process while the vertical retorts allowed continuous production of zinc. Why is continuous production an advantage over batch processes?

2. State two other advantages of the imperial smelting furnace (ISF) over the original Belgian horizontal retorts.

3. Write an equation for the reaction of zinc oxide with carbon monoxide.

4. Write an equation for the reaction of zinc with carbon dioxide.

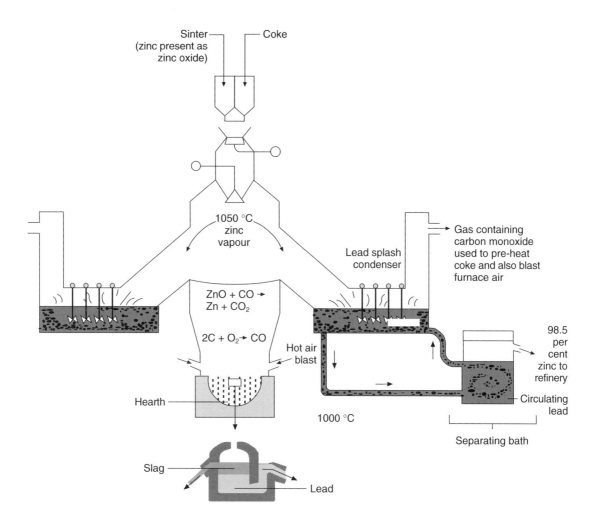

Derham's blast furnace

5. Some countries still operate the vertical retort first used in the 1930s and there are even some of the original Belgian horizontal retorts in use. Suggest a reason why these outdated processes are still used.

_____
_____
_____
_____

6. What might have been the consequences in World War II if the UK had not established its own zinc industry?

_____
_____
_____
_____

7. Zinc manufactured in the UK by the imperial smelting furnace (ISF) process is in competition with imported zinc for many uses. If imported zinc is cheaper what reasons are there to maintain the UK industry?

_____
_____
_____

8. Why was Avonmouth considered a good site for establishing the UK zinc industry?

_____
_____
_____
_____
_____

Uses of zinc

A Zinc for die-casting metal toys

The metal that was once used to make detailed models, for example of toy soldiers, was lead. Now, zinc-based die-casting alloys are used.

Here are some properties that make zinc alloys suitable.

- **Density** – Being a lot denser than plastic, metal toys have a pleasantly heavy feel.

- **Melting point** – Zinc melts at around 420 °C, similar to lead (328 °C), and much lower than copper or steel.

- **Fluidity** – When alloyed with a few per cent of magnesium and aluminium, zinc makes a molten metal which can be forced under pressure through a tiny hole into a steel die (like a mould) where it quickly cools and sets without shrinking.

- **Precision** – The finished toy is tough, strong and accurately shaped.

- **Corrosion resistant** – The finished toy will not rust away.

Questions

1. Which is very toxic – zinc, lead or plastic?

2. Which of these three materials cannot be given a high gloss paint finish?

3. Why is it useful to have a low melting point material for model making?

4. Why must molten zinc alloy cool and solidify quickly in the steel die?

B Zinc for brasses

Brass is an alloy (metal mixture) of copper and zinc. Rather than brass we ought to say brasses, because brass alloys can have anything between 20 per cent and 45 per cent zinc. The alloy gives a harder and stronger metal than zinc or copper alone, has high electrical conductivity, and is fairly corrosion resistant.

Brass is a good metal for ornamental use because it can be given such a high shine.

Questions

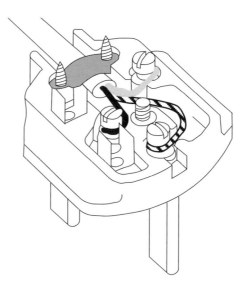

An electrical plug

5. Explain why brass is more suitable than copper for the pins on a plug, yet copper is used for the wiring.

   _____

   _____

   _____

6. Give two more places in the house where you often see brass, and suggest why brass is a suitable metal for each use.

   _____

   _____

C Zinc for making zinc oxide, ZnO

- Zinc oxide helps to keep your wheels bouncing along.
  Tyres contain about 5 per cent by weight of zinc oxide. Rubber in its natural state is a soft and stretchy material with long tangled chains of molecules. Zinc oxide is an important ingredient for the vulcanisation of rubber – the process that makes a tougher structure with more crosslinks between the molecules.

- Zinc oxide makes it all better.
  Plasters and ointments with zinc oxide in them help heal wounds. This is because zinc oxide prevents both bacteria and fungi from reproducing (it is called a bacteriostat and a fungistat).

- Zinc oxide is used as a sunscreen too, being a good absorber of ultraviolet light and a white, opaque compound. It has a low toxicity.

- Zinc oxide is also used in 'cattle-lick' because the cows are healthier if it is included in their diet. When zinc oxide is added to pig food it minimises skin warts.

Questions

7. Write down some properties of a material which would make a good tyre.

   _____
   _____
   _____

8. Why is zinc oxide used so much in nappy creams?

   _____
   _____

Zinc – for the roof over your head

The rain falling on a building must be drained away safely so that the wooden roof beams do not rot and the house below stays free from damp. Metal sheets are sometimes used as roofing material.

In Copenhagen they use copper

In London they use lead

In Paris they use zinc

Remember (or look up) the reactivity series.

1. Write down the three metals above in order of reactivity with the most reactive at the top of the list.

   _____

   _____

   _____

Zinc may seem too reactive to be left out in all weathers. In fact a chemical reaction occurs which takes place as it weathers, and protects it from further attack.

2. Find the colours of the minerals which are ores of the metals used as roofing materials and fill them in the spaces in the table.

Mineral	Chemical name	Formula	Colour
Sphalerite	Zinc sulfide	ZnS	
Galena	Lead sulfide	PbS	
Malachite	Hydrated copper(II) carbonate	$CuCO_3 \cdot Cu(OH)_2$	

Weathering – ageing well

3. The table below shows what happens to the different roofing metals as they weather in wind and rain. Fill in the last two columns. You may need a book of data, database or CD ROM.

Metal	Colour of roof when new	Colour after weathering (may take many years)	Colour of metal carbonate	Colour of metal oxide
Zinc	Shiny light grey	Dull bluish grey		
Lead	Shiny dark silvery grey	Dull whitish grey		
Copper	Shiny red-brown	Black, then green		

4. Explain what seems to be happening to the metals on the roofs.

   _____
   _____

5. Write word and balanced symbol equations for:

   a) the reaction of zinc with oxygen

   _____
   _____

   b) the reaction of copper(II) oxide (CuO) with carbon dioxide.

   _____
   _____

As it weathers, zinc slowly forms a thin layer of a mixture of zinc oxide, zinc hydroxide, zinc carbonate and zinc sulfide. All these compounds have the important properties of being insoluble in pure water and sticking strongly to the metal. This layer of weathered metal, called the patina, protects it from further attack. The same is true of copper and lead.

Why do metal roofs need replacing?

Metal roofing can deteriorate over the years for three main reasons.

1. Creep

Creep in lead roofing

A roof is under the stress of its own weight. Gradually that force can change the shape of the metal even though it stays in the solid state all the time. Lead, being so heavy and rather soft, is especially prone to this creep which is caused by layers of atoms sliding over one another.

2. Thermal expansion stress cracking

In the full glare of the sun, roofs warm up quickly and then cool by night. As a result, they expand and contract in daily cycles. This may cause stress cracks, or weaken the fixings which hold the metal onto the roofs. Alloying zinc with tiny percentages of copper and titanium reduces its coefficient of thermal expansion. This means that it expands and contracts less for a given temperature difference and so the thermal stresses are reduced.

3. Chemical weathering

Acid rain means that sulfuric and nitric acids are present in rain. These acids will react with the patina and form soluble sulfates and nitrates. Then the fresh metal below is also attacked.

Questions

1. What problems are caused by metal creep?

    _____

2. In what sort of climate would stress cracking be the greatest problem?

    _____

3. Rewrite the equations which show how the components of the patina react with sulfuric acid, filling in the missing words and formulae.

    a) zinc oxide + sulfuric acid → ——— ——— + water

    $ZnO(s)$ + ——— (aq) → $ZnSO_4(aq)$ + ——— (l)

    b) zinc hydroxide + sulfuric acid → ——— ——— + water

    $Zn(OH)_2(s)$ + ——— → ——— + 2——— (l)

c) zinc carbonate + sulfuric acid → ———— ———— + ———— ————
      + water

   ZnCO$_3$ + ———— (aq) → ————(aq) +———— (g) + ————(l)

4. Write word and balanced symbol equations for the reactions of the following compounds with nitric acid

   a) zinc oxide,

   _____
   _____

   b) zinc hydroxide

   _____
   _____

   c) zinc carbonate

   _____
   _____

The different ways of coating steel with zinc

There are several ways to put a zinc coat on steel. Zinc is expensive, nearly eight times more expensive than steel, so the method is chosen to suit the cost of the object.

Hot dip galvanising

The whole steel object is plunged into a bath of molten zinc at 460 °C for several minutes. The zinc does more than just coat the steel. It forms strong metallic bonds where iron and zinc atoms change places in their structures. This makes the coat into an alloy, which is harder than either metal on its own. Galvanising is costly, but puts on a thickish layer which is useful for more valuable objects. However, modern techniques for making continuous sheets of galvanised steel reduce the thickness of the zinc coating, and the use of special zinc alloys reduces the thickness still further.

Zinc dust painting

This is the simplest method of all and it can be done by anyone. The dust is in a solvent which evaporates away, leaving the zinc coat.

It is easy and cheap to do but tricky if shapes are intricate.

Zinc spraying

A gas flame blows molten zinc onto the object, which can be any size. A really thick layer can be built up by spraying several times. This is suitable for something which needs to last for many years and which may be hard to respray.

Electroplating

The cathode is the steel object to be plated and the anode is pure zinc. This is good for smaller objects because the zinc layer is thin but very even.

Sherardizing

Hot but not quite molten zinc dust at 375 °C is rolled around in a box with clean sand and the object to be plated. A thin, not very even coat of metal sticks to the steel object even if it has a very intricate shape. This is a fast and relatively cheap method of protection.

Questions

1. A typical use for zinc-coated iron is for a climbing frame in a playground. The end product must be hard, smooth and long lasting. Which method(s) of coating are most suitable for this use and why?

   _____

   _____

The different ways of coating steel with zinc: page 1 of 2

2. Which method is most suitable for coating nuts, bolts, nails and screws. Explain your answer.

   _____
   _____

3. The Forth road bridge in Scotland is made of steel. Why is spraying with zinc suitable to protect it?

   _____
   _____

4. Which method is best for treating a rust spot on a car?

   _____
   _____

Part 2 Practical work

Teacher's notes

Using carbon to extract copper from copper oxide could be used as an introduction to extracting less-reactive metals by displacement.

Zinc to the rescue consists of a set of four simple experiments to:

- find the place of zinc in the reactivity series; and
- use zinc as a sacrifical anode to prevent rusting.

Zinc in cells and batteries introduces the idea of a chemical cell.

Apparatus (per group)

Using carbon to extract copper from copper oxide
Each group of pupils will require:

Apparatus
- A crucible and lid
- Pipeclay triangle
- Tripod
- Bunsen burner
- Spatula.

Chemicals (per group)

- A spatula of powdered copper(II) oxide and of powdered carbon. The exact quantities are not critical.

  Note. the carbon in the charcoal reduces the black copper oxide to reddish-brown copper. The lid must not be removed until the crucible is cool or the hot copper will be re-oxidised by air.

Zinc to the rescue
Experiment 1
To compare the reactivity of zinc with some other metals

Each group of students requires:

Apparatus
- Three test-tubes with stoppers,
- Test-tube rack.

Chemicals
- A few small pieces of cleaned iron (or mild steel), zinc and copper sheet
- A spatula of drying agent (*eg* anhydrous calcium chloride granules)
- A few grams of sodium chloride to make a solution.

Zinc

Experiment 2
To compare how the metals react with acid
Each group of students requires:

Apparatus
- Three test-tubes
- Test-tube rack
- Spatula.

Chemicals
- A small spatula of each of powdered zinc, iron, and copper
- Wooden splints
- Approximately 100 cm^3 of 1 mol dm^{-3} sulfuric acid solution.

Experiment 3
Displacing metals
Each group of students requires:

Apparatus
- Two boiling tubes
- Spatula
- Thermometer (0–100 °C).

Chemicals
- Approximately 50 cm^3 of 1 mol dm^{-3} copper(II) sulfate solution,
- A spatula of powdered zinc and of powdered iron.

Experiment 4
Preventing rusting
Each group of students requires:

Apparatus
- Four test-tubes
- Test-tube rack
- Emery paper or glasspaper
- A pair of pliers.

Chemicals
- Steel wool,
- Potassium hexacyanoferrate(III) solution (50 cm^3 of 0.1 mol dm^{-3})
- A few drops of dilute (approximately 1 mol dm^{-3}) sulfuric acid solution
- Pieces of zinc, copper foil and magnesium ribbon.

Summing up – no apparatus needed

Zinc in cells and batteries

Each group of students requires:

Apparatus
- One 100 cm^3 beaker
- One voltmeter (0–3 V)
- Two crocodile clips
- Two leads
- One 1.5 V bulb
- One pea bulb (a torch bulb of the lowest voltage available)
- Emery paper.

Chemicals
- Strips of zinc, copper, lead, iron, aluminium, magnesium and a carbon electrode.

For the extension work, the cell will light a bulb if magnesium and copper are used, with 1 mol dm^{-3} hydrochloric acid as the electrolyte. A long piece of magnesium ribbon needs to be made into a concertina. The reaction is vigorous and it may be preferable to demonstrate this.

Using carbon to extract metals from their ores

Zinc needs a very high temperature to extract it from its ore but in the laboratory a similar process can be used to extract copper from copper(II) oxide.

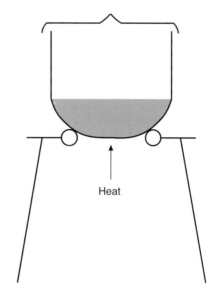

Heat

- Mix a spatula of black copper(II) oxide with a spatula of charcoal powder in a crucible and cover it with a lid. Heat it strongly on a tripod for about five minutes then allow it to cool to room temperature without lifting the lid.

- Examine the contents, looking carefully for any colour changes.

- Describe what has happened.

- Write word and balanced symbol equations for the chemical reaction that has taken place.

- Why is it important not to lift the lid before the crucible has cooled?

Zinc to the rescue

A layer of zinc on steel is a good way of preventing steel from rusting. Zinc does not rust even when in air and water, and it protects the steel underneath in a special way called a sacrificial method.

To compare the reactivity of zinc with some other metals

Experiment 1
To compare the resistance of zinc, iron and copper sheet to corrosion

Take a few small pieces of cleaned iron (or mild steel), zinc and copper sheet. Place one sample of each outdoors. Place the other samples in separate stoppered test-tubes containing:

a) a drying agent (*eg* anhydrous calcium chloride granules),

b) boiled water,

c) a small amount of salted water.

Leave the samples for about a week then examine them for signs of corrosion. Describe the appearance of each sample.

Experiment 2
To compare how the metals react with acid

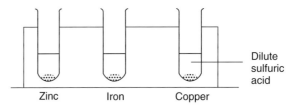

- Put three test-tubes into a test tube rack and 1/3 fill each with dilute (1 mol/dm^3) sulfuric acid.

- Add a small spatula measure of powdered metal to each test-tube as shown – zinc in the first, iron in the second, copper in the third. Leave the test-tubes in the rack, and if the mixture fizzes, trap the gas with your thumb and test it with a lighted splint.

- Write down your results and put the metals in order of reactivity, with the most reactive metal first.

Experiment 3 Displacing metals

In this experiment a metal is added to a solution of blue copper sulfate. If the metal is more reactive than copper it will displace the copper from the copper sulfate. The reaction gives out heat, and the greater the difference in reactivity, the more heat is given out.

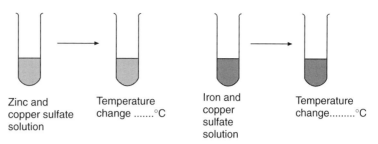

- 1/3 fill a test-tube with dilute(1 mol/dm³) copper(II) sulfate solution ($CuSO_4$). Record the temperature of the copper sulfate solution.
- Add a spatula measure of zinc powder and gently stir with the thermometer. Note the temperature difference.
- Repeat the experiment with iron powder in place of zinc powder.
- Write down your results and say which metal displaced the copper more vigorously, zinc or iron.

Further work

- What name is given to reactions which give out heat?
- What can you see in the test-tube after the reaction?
- Write a word equation for each reaction
- Write a balanced symbol equation for each reaction.

Experiment 4 Preventing rusting

In this experiment you are going to find out if the rate of rusting of iron can be changed by pairing it with other metals. (You will use a special corrosion indicator, potassium hexacyanoferrate(III) solution, which turns the water blue where rust is forming.)

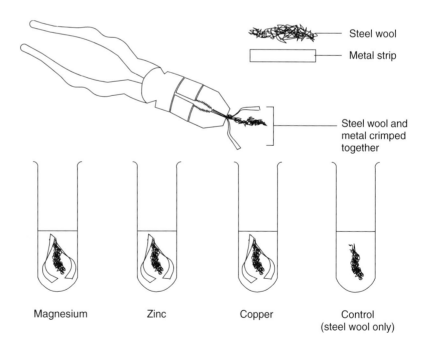

- Take some steel wool and divide it into four loose cigar-shaped tufts.
- Put a few drops of dilute sulfuric acid into 50 cm^3 of 0.1 mol/dm^3 potassium hexcyanoferrate(III) solution.
- Use this solution to half fill four test-tubes.
- Cut some pieces of zinc and copper foil and magnesium ribbon roughly the same width and length as the tufts and scratch them clean with some emery paper.
- Crimp the foil onto the steel wool with a pair of pliers to make a V shape. Push it down into the bottom of the test-tubes. Place some steel wool on its own in another test-tube.
- Watch the test-tubes carefully and arrange them in order of how quickly they go blue. Write down the order, with the fastest first.

Why use steel wool rather than a steel nail?

Why did you need to have steel wool on its own included in your trials?

Can you see a pattern in your results?

Summing up

When in contact with air and water, the surface layer of zinc atoms react to form zinc oxide and hydroxide. These are both insoluble solids which stick tightly to the metal below and clog any pits and scratches where air and water could get in. (Do you know another grey-looking metal which does the same thing? Think of saucepans.)

The iron in steel rusts steadily when in contact with water and dissolved air. If iron is attached to a less reactive metal, such as copper, it rusts faster than normal. If it is attached to a more reactive metal, such as zinc, is rusts more slowly, even when the iron is in contact with water and air. How does this work?

When iron is paired with zinc it protects the steel from rusting. Because zinc is more reactive than iron, zinc combines with oxygen (or anything else) in preference to iron.

The technical term for this is that zinc is acting as a sacrificial anode. Metals form ions as they react. Iron forms Fe^{3+} ions and zinc Zn^{2+} ions and in doing so they lose electrons. Since zinc is more reactive than iron it reacts more readily and, when forming the zinc ion Zn^{2+}, deposits its electrons onto the iron. This means that the iron atoms do not lose any electrons and iron stays as a metal.

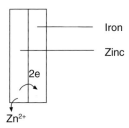

Zinc in cells and batteries

Some technical terms.

A cell is the basis of the batteries that we use for example in our portable radios and CD players. A battery is two or more cells connected together.

The total amount of charge you can get from a battery depends on how big it is, and the amount of metals and other chemicals it contains. The **current** shows how much charge can flow in a second.

The **voltage** provides the push that moves the electrons when the cells are doing their work. This is related to the reactivity of the elements in the cell.

Experiment: making a cell
You will be provided with strips of zinc, copper, lead, iron, aluminium and magnesium, and a carbon electrode. Clean the metal strips carefully with emery paper.

Set up the zinc and the other metal, connected with a voltmeter as shown. This is the structure of a simple cell – two different metals, connected with a wire through which electrons flow, to give a potential difference and a current. Fill in the table for the voltage.

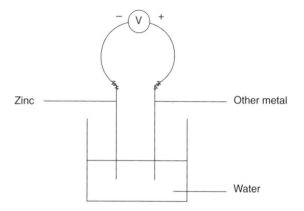

Element paired with zinc	Voltage/V
Copper	
Lead	
Iron	
Aluminium	
Magnesium	
Carbon	

- Write down the pair that gives the greatest voltage.
- Name a metal that might give a bigger voltage and try it if you can.
- What is different about magnesium paired with zinc?

Now try lighting a) a 1.5 V bulb b) a pea bulb with the cells that gave the best result. To light a bulb you need a relatively large current.

■ What is happening here?

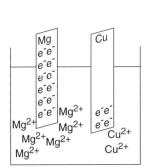 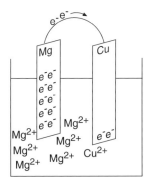

When two different metals are joined together with a metal wire as shown, an electric cell is created.

Metals can form ions as they react. Magnesium forms Mg^{2+} ions and copper forms Cu^{2+} ions and in doing so they lose electrons. Magnesium is more reactive than copper and reacts more readily. When magnesium and copper are joined together, the excess of electrons from the magnesium pass through the wire to the copper. While this is happening, an electric current flows.

Further work
To obtain a current that will light a bulb, the electron flow must be larger and keep going. This depends on a) the electrolyte b) the surface area of the metals.

Suggest an electrolyte that makes magnesium react faster. Predict how a change in surface area affects the current. Your teacher may ask you to try your ideas.

Find out how a dry cell works.

Part 3 Post-16

Teacher's notes

This comprehension exercise best follows from the information sheet *Zinc at Avonmouth* though it could stand alone. It revises aspects of enthalpy, entropy, and Le Chatelier's principle.

Answers

The thermodynamics of zinc extraction

1. a) $C(gr) + O_2(g) \rightarrow CO_2(g)$ (**NB** coke is not pure graphite.)

   b) $2C(gr) + O_2(g) \rightarrow 2CO(g)$

   or $CO_2(g) + C(gr) \rightarrow 2CO(g)$

2. Step 1 $ZnO(s) + CO(g) \rightarrow Zn(s) + CO_2(g)$

   $\Delta H = +64.8$ kJ mol^{-1} at 298 K

   Step 2 $\Delta G = \Delta H - T\Delta S$

   $= 64.8 - \dfrac{298 \times 14}{1000}$

   $= 64.8 - 4.2$

   $\Delta G = +60.6$ kJ mol^{-1}

   Therefore the reaction is not feasible.
   or
   $\Delta S_{total} = \Delta S_{syst} + \Delta S_{surr}$

   $\Delta S_{surr} = -\dfrac{\Delta H_{syst}}{T}$

   $= \dfrac{-64.8 \times 1000}{298}$

   $= -217.4$ J K^{-1} mol^{-1}

   $\Delta S_{tot} = +14.0 - 217.4 = -203.4$ J K^{-1} mol^{-1}

   Therefore the reaction is not feasible.

   b) $ZnO(s) + CO(g) \rightleftharpoons Zn(s) + CO_2(g)$ $\Delta H = +64.8$ kJ mol^{-1}

   Using Le Chatelier's Principle;
   Pressure – no effect (one molecule of gas on each side),
   Temperature – increasing temperature will force reaction to the right (endothermic reaction).

   c) the entropy change is small as there is one mole of solid and one mole of gas on each side of the equation.

Zinc

3. **a)** Assuming other values of ΔH_f are unchanged at this temperature

   $ZnO(s) + CO(g) \rightarrow Zn(g) + CO_2(g)$

   At 907 °C (1180 K) $\Delta H_f = +187.5$ kJ mol^{-1}

   b) $\qquad ZnO(s) + CO(g) \rightarrow Zn(g) + CO_2(g)$

   S / J K^{-1} mol^{-1} $\quad$ 43.6 $\quad$ 197.6 $\quad$ 362.6 $\quad$ 213.6

   $\Delta S_{syst} = +335.0$ J K^{-1} mol^{-1}

   The entropy change is much more positive than at 298 K as the zinc is now a gas (more disordered)

   c) $\Delta G = \Delta H - T\Delta S$

   $= 187.5 - \dfrac{1273 \times 335.0}{1000}$

   $= 187.5 - 426.5$

   $= -239.0$ kJ mol^{-1}

   The reaction is therefore feasible.

   or $\Delta S_{total} = \Delta S_{syst} + \Delta S_{surr}$

   $\Delta S_{surr} = -\dfrac{\Delta H_{syst}}{T}$

   $= \dfrac{-187.5 \times 1000}{1273}$

   $= -147.3$ J K^{-1} mol^{-1}

   $\Delta S_{total} = +335.0 - 147.3$

   $\Delta S_{total} = +187.7$ J K^{-1} mol^{-1}

   The reaction is therefore feasible

   d) A decrease in temperature could result in the zinc condensing to a liquid. Zn(l) will have a smaller value of entropy and this will reduce the value of ΔS_{total} so that the reaction may no longer be feasible. Also removal of zinc as a vapour will pull the equilibrium to the right (Le Chatelier's Principle).

4. **a)** Zinc will be liquid (it is between its melting and boiling temperatures).

   b) Zinc will be in the upper layer as it less dense.

The thermodynamics of zinc extraction

The feasibilty of a reaction can be determined using either the Gibbs free energy change, ΔG, or the total entropy change, ΔS_{total}. These are both temperature-dependent thermodynamic functions.

Zinc has an unusually low boiling point for a metal (907 °C) and this is a crucial and rather unusual property which makes the thermodynamics of the reduction process favourable. It is also important for the separating procedure (invented at Avonmouth) for purifying the zinc. Answering the questions below will let you investigate the feasibility of the reaction.

The chemistry in a zinc blast furnace is similar to that in the extraction of iron in a blast furnace. Coke (carbon) acts as a fuel and also generates carbon monoxide.

1. Write two balanced equations to represent

    (a) the complete combustion of carbon and

    (b) the production of carbon monoxide

2. The main reduction reactions are:

    $ZnO(s) + C(s) \rightarrow Zn(g) + CO(g)$

    $ZnO(s) + CO(g) \rightarrow Zn(g) + CO_2(g)$

    At room temperature (298 K) the zinc would be a solid.

    a) Using steps 1 and 2, show that the reaction:

    $ZnO(s) + CO(g) \rightarrow Zn(s) + CO_2(g)$

    would not be feasible at this temperature.

    Step 1 Use the enthalpy of formation values to calculate an enthalpy change for this reaction at 298 K.

	ZnO	CO	CO_2
$\Delta H_f^\ominus$ / kJ mol^{-1}	–348.3	–110.0	–393.5

    Step 2 The standard entropy change for this reaction at 298 K is +14.0 JK^{-1} mol^{-1}. Establish the feasibility of the reaction at 298 K by calculating either the Gibbs Free Energy change ($\Delta G = \Delta H - T\Delta S$) or the total entropy change (ΔS_{total}).

    b) Use Le Chatelier's principle to comment on the effect on yield of increasing (i) temperature and (ii) pressure on the above reaction.

    c) Comment on why the entropy change is a small value.

3. Above 907 °C the zinc boils so the reaction becomes

    $ZnO(s) + CO(g) \rightarrow Zn(g) + CO_2(g)$

    (a) Using the additional ΔH_f value for Zn(g) of +122.7 kJ mol^{-1} calculate a new value for the enthalpy change of this reaction above 907 °C.

(b) Use the following standard entropy values to calculate the entropy change for this reaction and comment on your answer.

	ZnO(s)	+ CO(g)	→	Zn(g)	+	CO_2(g)
S / $JK^{-1}mol^{-1}$	43.6	197.6		362.6		213.6

(c) Determine the feasibility of this reaction at 1000 °C (1273 K) by calculating ΔG or ΔS_{total}.

(d) As the gases leave the furnace the mixture can cool unless heated. Use Le Chatelier's Principle to explain why cooling needs to be avoided and what effect any decrease in pressure would have.

4. The melting, boiling points and densities of zinc and lead are as follows:

	Melting point /°C	Boiling point /°C	Density(solid) / g cm^{-3}
Zinc	420	903	7.14
Lead	328	1740	11.34

The zinc is removed quickly by quenching the mixture in molten lead at about 550 °C.

(a) What state will the zinc be in at a temperature of 550 °C?

(b) At the initial high temperatures the zinc dissolves in the molten lead, but as the mixture moves through the process it cools and the metals separate into layers. Explain whether zinc will be the upper or lower layer.

Zirconia

Introduction

This booklet is the result of a Learning Material Workshop organised by The Royal Society of Chemistry in conjunction with The Institute of Materials and The Worshipful Company of Armourers and Brasiers. A group of chemistry teachers spent the day at MEL Chemicals, Bolton, which manufactures zirconia. The day included a presentation by the company and tour of the plant. The following day was spent brainstorming and drafting the material which is presented here in edited form.

Acknowledgement

The Royal Society of Chemistry thanks MEL Chemicals and in particular Dr Ian McAlpine, who gave freely of his time and expertise both during the workshop and afterwards. The Society is also grateful to Professor Ron Stevens of The University of Bath, Dr David Moore of St Edward's School, Oxford and Dr Frank Ellis of GlaxoWellcome, Stevenage who read drafts of the material.

The teachers who wrote this material are:

David Cooper; Sutton Valence School, Maidstone, Kent

Gina Golledge; Waltham Forest College, Walthamstow, London

The material consists of:

- teacher's background notes on zirconium and some of its compounds, including answers to the comprehension exercise;

- some suggestions for searching the internet to find out about zirconium and its compounds which could be used to begin an information and communication technology (ICT)-linked project; and

- a comprehension exercise for post-16 students.

While zirconium itself does not figure in post-16 syllabuses, the comprehension exercise uses the element and its chemistry as a vehicle to cover some familiar chemistry topics – oxidation states, coordination numbers, electron configurations *etc* – in an unfamiliar context.

When reading this material, one needs to take care with nomenclature:

- 'zirconium' refers to the metallic element;

- 'zirconia' refers to zirconium(IV) oxide, ZrO_2; and

- 'zircon' refers to zirconium(IV) silicate – this is a compound oxide ($ZrO_2.SiO_2$) but is often represented $ZrSiO_4$

Teacher's notes

Zirconium and zirconia

The name 'zirconium' comes from the Persian word *zargun* meaning 'gold-coloured'. Zirconium is a relatively unfamiliar metallic element, and yet its compounds (especially zirconium oxide, zirconia) have many everyday applications. Zircon (zirconium silicate, $ZrSiO_4$) has been known for centuries as a semi-precious stone but it was not until the 1920s that pure zirconium metal was first produced. Since then there has been a steady increase in the diversity of the uses of the metal and its compounds. One of the reasons why interest in zirconium continues to grow is that, unlike many of the other metals in common usage, zirconium has a low toxicity and is classified as being non-hazardous to the environment.

Zirconium is the 17th most abundant element on the Earth (more common than zinc, tin and mercury). It is found in igneous rocks, such as schists, gneiss, syenite and granite. In these rocks it exists in the form of baddeleyite which is zirconium oxide (zirconia, ZrO_2) which has iron, titanium and silicon oxide impurities. The most commercially important mineral is zircon ($ZrSiO_4$). This was originally associated with igneous deposits, but weathering and natural concentration due to its high density (4.6 g cm^{-3}) has produced large secondary deposits in beach sands. These important secondary deposits are in Australia, South Africa, Asia and the East coast of the US.

The process of obtaining pure zirconium compounds from the impure zircon mineral is complex, but can be represented by the flow diagram (*Fig 1*).

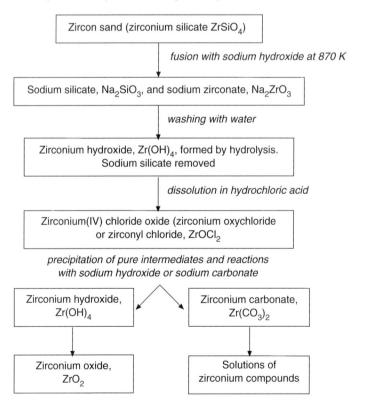

Figure 1 Flow Diagram

Zirconium data

Symbol	Zr
Atomic number	40
Relative atomic mass	91.22
Zirconium isotopes:	^{90}Zr, 51.5 per cent; ^{91}Zr, 11.2 per cent; ^{92}Zr, 17.1 per cent; ^{94}Zr, 17.4 per cent; ^{96}Zr, 2.8 per cent
Density	6.5 g cm^{-3}
Melting point /K	2125
Boiling point /K	4650
Electronic configuration	[Kr] 4d^{2}5s^2
Ionisation energies/kJ mol^{-1}	1st 660; 2nd 1267; 3rd 2218; 4th 3313; 5th 7860; 6th 9500; 7th 11 200; 8th 13 800
Electronegativity	1.4 (Pauling scale)
Metallic radius /nm	0.158
covalent radius /nm	0.145
Ionic radius nm	0.072
$\Delta H_f^\ominus$ (ZrO$_2$) kJ mol^{-1}	-1080
$\Delta G_f^\ominus$ (ZrO$_2$) kJ mol^{-1}	-1023
$S^\ominus$ (ZrO$_2$) J K^{-1} mol^{-1}	50.3

Zirconia crystals

Zirconia exists in three different crystalline forms – monoclinic, tetragonal and cubic. In the monoclinic phase the Zr^{4+} ion has seven-fold coordination (*ie* it is surrounded by seven oxygens); in the tetragonal phase and cubic phase the Zr^{4+} ion has eight-fold coordination.

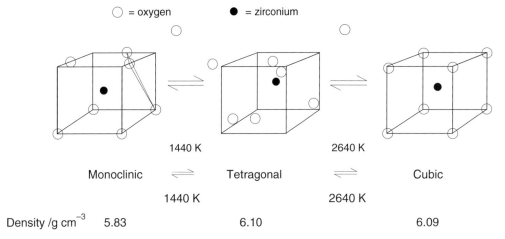

	Monoclinic	⇌ 1440 K	Tetragonal	⇌ 2640 K	Cubic
Density /g cm^{-3}	5.83		6.10		6.09

When the tetragonal phase transforms to the monoclinic phase – on cooling – the volume actually increases by about 4 per cent and zirconia is brittle over such temperature changes.

By adding a few per cent of oxides of metals such as magnesium, calcium and yttrium, makes the cubic form of zirconia stable from zirconia's melting point (2950 K) down to room temperature. This effect gives zirconia ceramics their unusual and highly valued properties. This is a complex effect but in simple terms, ions of the added metals take the place of some of the zirconium ions in the zirconia crystal lattice. However, these oxides are oxygen deficient compared with zirconia – the formulae are MgO, CaO and Y_2O_3 respectively compared with ZrO_2). Their presence distorts the shape of the monoclinic and tetragonal lattices and has the effect of making the cubic form the most stable form down to room temperature.

Uses of zirconium and its compounds

Zirconium metal

- Zirconium is used for the cans that hold reactor fuel rods in the nuclear industry.

- It is used as pressure tubes in Canadian nuclear reactors and in the reactors of the US Navy's nuclear submarines.

- Zirconium is used as an ingredient to increase the strength of magnesium alloys. For example, when such alloys are used in aircraft, this gives lighter weight components for the same strength. This in turn leads to more efficient use of fuel and consequent reduction of air pollution.

Zirconium oxide

Ceramics

Zirconia can be used to make ceramics, and it is this use of zirconia which has most excited scientists over the past few years.

Ceramics have been used for thousands of years for making vases, tiles *etc*. The first ceramic was probably discovered by accident - possibly by noticing that lumps of soft clay become extremely hard when they are left to dry and then fired. The drawback with traditional ceramics is that they are brittle, but zirconia ceramics suffer less from this disadvantage. In fact zirconia ceramics have remarkable properties of strength, hardness and wear resistance in addition to withstanding attack from molten metal, organic solvents, acids and alkalis. They can also withstand high temperatures for long periods even under severe mechanical stress.

One of the new applications is knives and scissors where the zirconia ceramic can be engineered to produce extremely fine and sharp blades which have a hardness of about 9 on the Mohs' scale (on which diamond has a value of 10). This is significantly harder than steels, and of course, ceramics do not rust. These implements give an exceptionally good, smooth cut in even the toughest of uses – they can be used to cut through Kevlar® – as used in riot shields and bullet proof vests!

High performance scissors and knives are produced for everyday use, and these are also in great demand by deep sea divers who require sharp blades which will not corrode in marine environments. Zirconia ceramic blades flex as well as steel and are non-magnetic, anti-static, and don't cause any metallic contamination. A new type of golf club head has been manufactured from zirconia ceramics where the extreme hardness enables a crisper and harder driving force to be delivered to the ball.

As a result of their strength, hardness and other properties, zirconia ceramics are being considered for a huge range of industrial applications which include motor engine components, high speed cutting tools, heat resistant linings in furnaces, containers for molten metals, and heat shields for space vehicles.

Some other uses of zirconia include:

- cubic zirconia gemstones ('fake diamonds') – the optical properties of cubic zirconia are superior to those of diamond;

- catalysts: zirconia is not only used as a catalyst in its own right, but also as a support medium and enhancer for other catalysts. This enables ctalysts to be used at higher temperatures or under severe conditions. Catalytic converters in vehicles contain zirconia;

- ceramic colours: made by adding compounds of other transition metals to zirconia, used in ceramic tiles and sanitary ware – baths, wash basins and toilet bowls – that can replace lead in paint. Monoclinic zirconia is used here.

- electroceramics used in piezoelectrics – gas lighters, *etc* – and capacitors. (Zirconia has some rather peculiar electrical properties and can under certain circumstances become electrically conducting.); and

- solid electrolytes: used in fuel cells and in oxygen sensors used in combustion control systems in boilers and in some car engines.

Other zirconium compounds

Zirconium phosphate is used in the ion-exchange medium in kidney dialysis machines

Zirconium on the internet

Further information on zirconium chemistry in general and MEL Chemicals in particular is available on the internet.

MEL Chemicals has a web site at www.zrchem.com

This site contains further information about the company as well as more specialised detail about the uses of zirconium and zirconia.

Maniago is an Italian company which produces, amongst other things, divers' knives made from zirconia. Their web site is www.italpro.com.

Other information about zirconia and related matters can be found by using a search engine.

An *Infoseek* search for information on the keyword Zirconia produces well over 1000 references, so a narrower more precise search instruction is more useful. For example, searching for '+zirconia +ceramics -cubic' gives about a dozen hits. (This searches for documents which contains the words zirconia and ceramics, but omits any which contain the word cubic). A search for '+zirconia +cubic +jewellery' gives information only on the uses of cubic zirconia as a gemstone and results in about ten sources.

Other search engines may work slightly differently and this could be a good teaching point. Searches like this could form the basis of a ICT exercise on using the internet to find information, and should develop ICT skills as well as turn up some interesting chemistry.

Answers

1. Fuse means to melt the two solids together.

2. Hydrolysis means reaction with water.

3. $ZrSiO_4 + 4NaOH \rightarrow Na_2ZrO_3 + Na_2SiO_3 + 2H_2O$

   $Na_2ZrO_3 + 3H_2O \rightarrow Zr(OH)_4 + 2NaOH$

   $Zr(OH)_4 + 2HCl + ZrOCl_2 + 3H_2O$

4. The sodium silicate dissolves in water.

5. + IV

6. Zr(IV) has lost the two 5d electrons and the two 6s electrons and is therefore $5d^0$. Electronic transitions between part-filled d-orbitals are what causes the colour of compounds of d-block elements. (A d^0 electron arrangement is why Ti(IV) compounds are colourless and a d^{10} arrangement is why Zn(II) does not behave as a transition metal.)

7. 74 per cent

8. 72.6 kg

9. a) $[Xe] 4f^{14} 5d^2 6s^2$

   b) [Kr] refers to the electron configuration of the inert gas krypton
   ie $1s^2 2s^2 2p^6 3s^2 3p^6 3d^{10} 4s^2 4p^6$

   c) The 4f electrons shield only poorly the outer electrons from the nuclear charge because of their shape. (This is called the lanthanide contraction.)

   d) The whole outer shell of electrons has been lost in the +IV state.

10. Hf^{4+} and Zr^{4+} have almost identical ionic radii. Therefore their oxides have very similar properties making them very difficult (and therefore expensive) to separate.

11. a) At the corners of a cube of which the zirconium ions is at the centre.

    b) 6+. Each Zr is $^{4+}$ and there are 2 OH^-

    c) 7

12. Relative atomic mass of zirconium = 91.3

Student material

Zirconium and its compounds

Zirconium – atomic number 40 – is an element in the d-block of the Periodic Table. Its main industrial uses are as the oxide (ZrO_2) usually called zirconia. Purified zirconia is used to produce gemstones similar to diamonds but much cheaper. The other major use of the oxide is in making tough, heat resistant ceramics, used in engine components, cutting tools, knives and golf clubs. Recently zirconia has been used to improve the efficiency of catalytic converters in cars. When mixed with compounds of other d-block elements, such as iron, vanadium and cobalt, the oxide forms pigments used for colouring washbasins, baths and ceramic tiles. The colours are heat resistant and do not fade, because the added coloured ions lock permanently into the oxide lattice.

Zirconium ore comes from Australia, Asia, South Africa. and the East coast of the US. It is called 'beach sand', because it also contains silicon dioxide (SiO_2) and has the formula $ZrO_2.SiO_2$. often written as $ZrSiO_4$. The zirconium oxide is extracted from the ore in a series of steps. The first stage is to fuse the ore with sodium hydroxide to form sodium zirconate (Na_2ZrO_3), and sodium silicate (Na_2SiO_3). The mix is then washed with water and the zirconate hydrolyses to form complex hydrated zirconium hydroxides. Acid is added to the mix and zirconium salts form, which can be precipitated. The zirconium salts can then undergo further reaction to form zirconium oxide.

The process is summarised in the flow diagram.

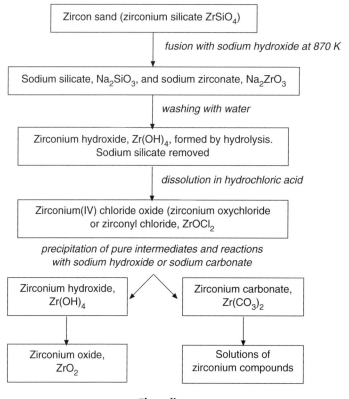

Flow diagram

One of the complex zirconium hydroxides is thought to contain the ion below

$$\left[\begin{array}{c} \text{structure with two } Zr^{4+} \text{ centres, each coordinated by H}_2\text{O ligands, bridged by two OH groups} \end{array} \right]^{4+}$$

Zirconium hydroxide complex ion

This ion can be dehydrated to give the bridged structure below.

$$\left[\begin{array}{c} \text{structure with two Zr centres coordinated by H}_2\text{O ligands, bridged by a single O} \end{array} \right]^{6+}$$

Bridged structure

After purification the final powder is 98 per cent ZrO_2 and 2 per cent HfO_2, hafnium oxide. The hafnium oxide is not normally separated from the zirconia.

Hafnium and titanium are elements in the same group as zirconium in the d-block of the Periodic Table. Some of their properties are given in the table.
The electron configuration of hafnium has been omitted.

Property	Titanium	Zirconium	Hafnium
Atomic number	22	40	72
Electronic configuration	$[Ar]3d^24s^2$	$[Kr]4d^25s^2$	$[Xe]$
Atomic radius (nm)	0.147	0.160	0.158
Radius in +IV oxidation state /nm	0.061	0.072	0.071
Density /g cm^{-3}	4.5	6.51	13.28

Some properties of titanium zirconium and hafnium

Questions

1. What does 'fuse the ore with sodium hydroxide" mean?

2. Explain the term 'hydrolysis'.

3. Write balanced equations for the reactions in the flow diagram by which the $ZrSiO_4$ in beach sand is converted into $ZrOCl_2$. There are three steps.

4. Suggest how the sodium silicate is removed during the extraction process.

5. What is the oxidation state of the zirconium in sodium zirconate Na_2ZrO_3?

6. Most compounds of the d-block elements are coloured. Given that the electronic configuration of Zr is $[Kr]4d^25s^2$, explain why Na_2ZrO_3 is colourless.

7. The price of ZrO_2 depends on the total mass of zirconium per 100 kg in the final powder. Calculate the percentage of Zr by mass in the oxide.
   (A_rs: Zr = 91.22, O = 16.0)

8. What mass of zirconium is there in a 100 kg batch of 98 per cent pure zirconium oxide,?

9. Use the information in the table to answer the following questions.
   a) Write down the rest of the electronic configuration for Hafnium.
   b) Explain what is meant by [Kr] in the electronic structure of zirconium as given in the table.
   c) Suggest a reason for the similarity in radii between zirconium and hafnium despite the difference in atomic number.
   d) Explain why the radius of zirconium in the +IV oxidation state is so much smaller than its atomic radius.

10. Many properties of the elements are governed by the charge/size ratio. Suggest why the impurity in hafnium oxide is not removed from the zirconium oxide by the chemical manufacturer.

11. a) Suggest the geometry of the arrangement of the groups surrounding each Zr atom in the ion below.

    $$\left[\begin{array}{c} H_2O \quad OH_2 \quad \overset{H}{\underset{|}{O}} \quad H_2O \quad OH_2 \\ H_2O\text{---}Zr^{4+}\text{---}\text{---}Zr^{4+}\text{---}OH_2 \\ H_2O \quad OH_2 \quad \underset{|}{\overset{|}{O}} \quad H_2O \quad OH_2 \\ H \end{array} \right]^{n+}$$

    b) What is the charge on this ion?
    c) What is the coordination number of each Zr atom in the bridged structure?

12. Naturally-occurring zirconium consists of the following isotopes:

^{90}Zr 51.5 per cent
^{91}Zr 11.2 per cent
^{92}Zr 17.1 per cent
^{94}Zr 17.4 per cent
^{96}Zr 2.8 per cent

Use these figures to calculate the relative atomic mass of zirconium to three significant figures.